International Conference on Mountain Hydrology and Cryosphere

Proceedings of the International Association of Hydrological Sciences (PIAHS) Volume 387, 2024

Kathmandu and Dhulikhel, Nepal
9-10 November 2023

Editors:

Rijan Bhakta Kayastha
Hari Krishna Shrestha
Dhiraj Pradhananga

ISBN: 979-8-3313-1936-6

Printed from e-media with permission by:

Curran Associates, Inc.
57 Morehouse Lane
Red Hook, NY 12571

Some format issues inherent in the e-media version may also appear in this print version.

Printed with permission by Curran Associates, Inc. (2025)

International Association of Hydrological Sciences
Centre for Ecology and Hydrology
Wallingford, Oxfordshire OX10 8BB
United Kingdom

Phone: 441 491 692 442
Fax: 441 491 692 448

http://iahs.info/

Additional copies of this publication are available from:

Curran Associates, Inc.
57 Morehouse Lane
Red Hook, NY 12571 USA
Phone: 845-758-0400
Fax: 845-758-2633
Email: curran@proceedings.com
Web: www.proceedings.com

TABLE OF CONTENTS

Proc. IAHS, 387, 1–2, 2024
https://doi.org/10.5194/piahs-387-1-2024

Preface: Mountain Hydrology and Cryosphere

Rijan Bhakta Kayastha[1], **Hari Krishna Shrestha**[2,3], **and Dhiraj Pradhananga**[4,5]

[1]Himalayan Cryosphere, Climate and Disaster Research Center, Department of Environmental Science and Engineering, School of Science, Kathmandu University, Dhulikhel, Nepal
[2]Society of Hydrologists and Meteorologists (SOHAM), Kathmandu, Nepal
[3]Department of Civil Engineering, Nepal Engineering College, Bhaktapur, Nepal
[4]Department of Meteorology, Tri-Chandra Multiple Campus, Tribhuvan University, Kathmandu, Nepal
[5]The Small Earth Nepal, Kathmandu, Nepal

Correspondence: Rijan Bhakta Kayastha (rijan@ku.edu.np)

Published: 18 November 2024

Mountain regions, with their complex terrain and intricate interplay of climate, cryosphere, and hydrology, are vital for the wellbeing of millions of people globally. These regions are governed by unique hydrological systems and the cryosphere, primarily governed by the climate dynamics in these regions, with snowmelt and glacier ice melt playing a critical role. However, climate change is increasingly impacting these systems, across mountain regions such as the Himalayas, Karakoram, Hindu-Kush, Pamir, Tian Shan, Andes, Alaska Range, Caucasus, Alps, Transantarctic Mountains, and the Rockies. Compared to a few decades ago, the patterns of precipitation, particularly monsoon dynamics, and snowmelt and glacier melt have shifted rapidly in these ranges.

In south Asia, the Asian monsoon and the processes of snowmelt and glacier melt are pivotal to the national economies. The monsoon is directly linked to agricultural productivity and disaster occurrences; a favorable monsoon results in a robust national economy and fewer casualties due to reduced incidents of floods and landslides. Similarly, snowmelt and glacier melt serve as crucial water sources during low-flow periods, especially in the pre-monsoon season, supporting hydropower production and other essential uses. Therefore, systematic studies of these rivers and diligent monitoring of snow, glaciers, and glacial lakes are indispensable in mountain regions.

To address the multifaceted challenges and opportunities related to mountain hydrology and the cryosphere, the International Conference on Mountain Hydrology and Cryosphere (ICMHC) was held in Kathmandu and Dhulikhel on 9–10 November 2023. Participated by 147 researchers, scientists, students, engineers, planners, policymakers, and developers and experts from 17 countries, the conference featured 10 technical and 3 formal sessions, encompassing 75 oral and 22 poster presentations. The topics covered were diverse, including snow hydrology in a changing climate, hydrological modeling, and water quality. Additionally, discussions extended to transboundary water resources; hydro-meteorology in agriculture; the cryosphere's impact on water resources, tourism, and society; glacier mass balance; debris-covered glaciers; permafrost hydrology; and water-induced disasters such as avalanches, glacier lake outburst floods (GLOFs), landslides, and landslide dam outburst floods (LDOFs). This forum enabled participants to exchange perspectives; share innovative research; and gain new insights into cryospheric changes, hydrological modeling, and climate change impacts.

Following the conference, presenters were invited to submit their full papers for these proceedings, entitled "Mountain Hydrology and Cryosphere". A total of 15 papers were accepted, reviewed, and published in this issue, covering most of the topics discussed at ICMHC 2023. We extend our gratitude to the authors, reviewers, and editors, whose contributions made these proceedings possible.

Editors of this issue

Rijan Bhakta Kayastha, national correspondent to the IAHS from Nepal and Kathmandu University (KU), Nepal

Hari Krishna Shrestha, Society of Hydrologists and Meteorologists (SOHAM) Nepal, and Nepal Engineering College, Nepal

Dhiraj Pradhananga, The Small Earth Nepal (SEN) and Tribhuvan University, Nepal

Reviewers of this issue

Madan Lall Shrestha, advisor to The Small Earth Nepal and Nepal Academy of Science and Technology, Nepal

Rijan Bhakta Kayastha, national correspondent to the IAHS from Nepal and Kathmandu University (KU), Nepal

Hari Krishna Shrestha, Society of Hydrologists and Meteorologists (SOHAM) Nepal and Nepal Engineering College, Nepal

Dhiraj Pradhananga, The Small Earth Nepal (SEN) and Tribhuvan University, Nepal

Ranjan Kumar Dahal, Tribhuvan University, Nepal

Bijay Thapa, Kathmandu University, Nepal

Disclaimer. Publisher's note: Copernicus Publications remains neutral with regard to jurisdictional claims made in the text, published maps, institutional affiliations, or any other geographical representation in this paper. While Copernicus Publications makes every effort to include appropriate place names, the final responsibility lies with the authors.

Acknowledgements. This issue of the proceedings of the International Conference on Mountain Hydrology and Cryosphere was made possible through the full support of the authors, reviewers, and editors. We acknowledge Copernicus Publications for their assistance in editing, typesetting, and compiling this issue. Special thanks go to the organizers of ICMHC 2023 – the IAHS, The Small Earth Nepal (SEN), and Kathmandu University – and the co-organizers – Tribhuvan University, the Department of Hydrology and Meteorology (DHM) of the Government of Nepal, the Nepal Tourism Board (NTB), the International Centre for Water Resources and Global Change (ICWRGC), the International Centre for Integrated Mountain Development (ICIMOD), the International Water Management Institute (IWMI), Global Water Futures (GWF) of the University of Saskatchewan, Society of Hydrologists and Meteorologists (SOHAM) – Nepal, and UNESCO. We also express our appreciation to the other associated partners of the conference: the Nepal Academy of Science and Technology (NAST); the Water Resources Research and Development Centre (WRRDC); the Center of Research for Environment, Energy and Water (CREEW); and Real-Time Solutions Pvt. Ltd. (RTS).

Proc. IAHS, 387, 3–8, 2024
https://doi.org/10.5194/piahs-387-3-2024

Open Access PIAHS

Assessing the characteristics of extreme floods in Nepal

Aabhash Bhattarai[1], **Utsav Bhattarai**[1,2], **Koshish Raj Maharjan**[1], **and Laxmi Prasad Devkota**[1,3]

[1]Water Modeling Solutions Pvt. Ltd. (WMS), Kathmandu, 44600, Nepal
[2]Institute for Life Sciences and the Environment, University of Southern Queensland, Toowoomba, Queensland 4350, Australia
[3]Nepal Academy of Science and Technology (NAST), Kathmandu, 44600, Nepal

Correspondence: Aabhash Bhattarai (er.aabhash21@gmail.com)

Published: 18 November 2024

Abstract. This study examines the characteristics (magnitudes, trends, and frequency of occurrences) of extreme floods in Nepal, a country that is at significant risk from floods. Daily discharge data from 1980 to 2015 of three gauging stations (Chisapani of Karnali Basin, Devghat of Narayani Basin, and Chatara of Koshi Basin) were used to assess the largest 1 % of flows, the annual top five high flows, and floods of different return periods (2-, 5-, 10-, 20-, and 100-year). In addition, temporal trend analysis of the flood peaks was carried out using the Mann–Kendall test and Sen's slope estimates. Results show that the magnitudes of the largest 1 % flows range from 6310 to 17 900, from 6967 to 12 100, and from 6080 to 9610 $m^3 s^{-1}$ at Chisapani, Devghat, and Chatara, respectively. The monsoon, especially from mid-June to early September, consistently witnesses over 90 % of 1 % extreme flows, with August registering more than 51 % of these occurrences. July and August combine for 81 % of the top five flow events, predominantly in August. Despite insignificant flow changes at a 95 % confidence level, extreme floods (2-, 5-, 10-, 20-, and 100-year return periods) are concentrated heavily in July and August, with August's second fortnight recording the most flood events. This assessment emphasizes July and August as critical months for extreme floods, aiding Nepalese authorities in planning dynamic resource allocation, disaster response, and effective flood management.

1 Introduction

Floods yield catastrophic consequences: casualties, ravaging of infrastructure and property, and uprooting of communities. Astonishingly, of the 11 072 disasters reported globally between 1970 and 2019, a staggering 44 % were attributed to floods leading to 330 000 deaths and economic losses worth USD 3.6 trillion (WMO, 2019). These figures underscore the urgent need to better comprehend flood behavior and implement effective flood management strategies, particularly in vulnerable developing countries.

In the context of Nepal, the recurring specter of floods has ingrained itself as a formidable natural challenge, wreaking extensive devastation throughout its topographically diverse landscape. Nepal is ranked 10th in the world for the highest percentage of its total gross domestic product (GDP) being affected by river floods each year (Luo et al., 2015). For example, the 2017 floods impacted 35 of the country's 77 districts and completely destroyed or partially damaged more than 190 000 houses, displaced tens of thousands of people, and made many homeless (National Planning Commission, 2017).

Past flood-related studies in Nepal delved into different aspects of flood disasters. For instance, Bhattarai et al. (2022) modeled the socioeconomic impacts of floods in the East Rapti River under climate change. Similarly, Devkota et al. (2013) used indicators to analyze people's perceived flood vulnerability in two southern districts. Moreover, the hydrological cycle has been altered due to land use change (Devkota et al., 2023). Thus, a collective analysis of the most extreme flood magnitudes together with their occurrences, temporal variation, and trends needs attention. This study attempts to contribute to these aspects.

While extensive research has focused on assessing the magnitudes of floods (Aryal et al., 2020; Devkota et al., 2020; Gautam and Dulal, 2013), there remains a notable gap

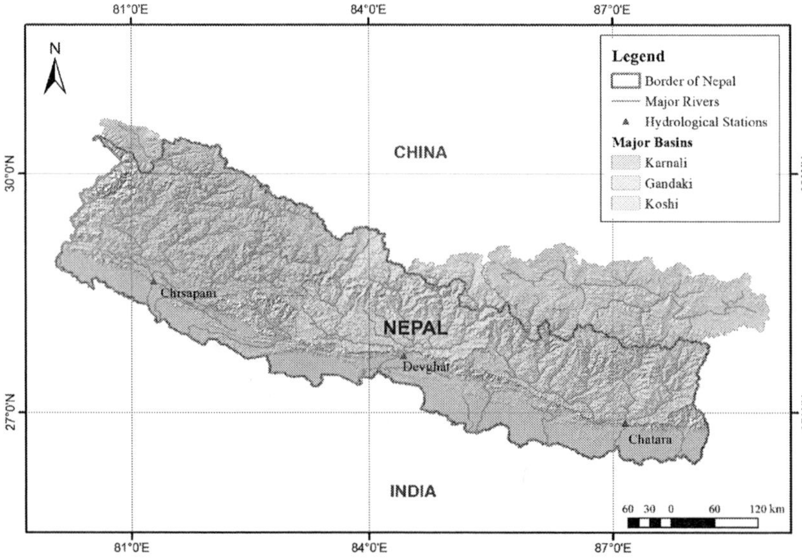

Figure 1. Location map of the study area.

in analyzing their temporal distribution. Similarly, most previous studies, such as Adhikari et al. (2023) and Khatiwada et al. (2018), were carried out in specific river basins. Hence, a comprehensive examination of the temporal flood behavior consisting of the major river basins of Nepal is needed. We focus our research on addressing these issues.

Therefore, this study seeks to conduct a comprehensive analysis of extreme flood events considering long-term hydrological data, focusing on the temporal distribution of extreme flood occurrences at the outlets of Nepal's three major river basins. Specifically, it aims to carry out trend analysis of the mean daily flood peaks and to analyze the temporal distribution of the top 1 % of extreme flows, the five highest annual flows, and floods of different return periods at fortnightly resolution.

2 Study area

This study was carried out in Nepal, considering the outlets of the three largest river basins: Karnali at Chisapani, Narayani at Devghat, and Koshi at Chatara (Fig. 1). Table 1 shows the characteristics of the study basins.

3 Methodology

Daily flow data from 1980 to 2015 of three hydrological stations (Chisapani, Devghat, and Chatara) were obtained from the Department of Hydrology and Meteorology (DHM), Government of Nepal. The top 1 % of the flows for each station were extracted for this duration. Similarly, the five highest flows from the 36 years were also extracted for each station. The counts of extreme flows across each fortnight were calculated. The contribution of each fortnight to the occurrences of these extreme flows was also quantified for the analysis of the temporal distribution of the top 1 % flows and the five highest flows. Trend analysis of the annual maximum of daily flows was carried out using the Mann–Kendall (MK) test and Sen's slope estimates. Since the trend analysis showed that changes in flows are insignificant at a 95 % confidence level, flood frequency analysis of the annual maximum of daily flows was carried out using the stationary approach based on Gumbel's extreme-value distribution method for flood magnitudes of 2-, 5-, 10-, 20-, and 100-year return periods for each of the three stations. The count of the flows exceeding the magnitudes of different return period floods was calculated. The contribution of each fortnight to the occurrences of these extreme flows was also quantified

Table 1. Characteristics of the study basins.

Basin (outlet)	Catchment area (km^2)	Average annual discharge ($m^3\,s^{-1}$)	Maximum discharge ($m^3\,s^{-1}$)
Karnali (Chisapani)	45 437	1380	17 900
Narayani (Devghat)	31 747	1514	12 100
Koshi (Chatara)	58 273	1526	9610

Table 2. Temporal distribution of the top 1 % extreme flows.

Month	Fortnight	Count of the top 1 % extreme flows		
		Karnali, Chisapani	Narayani, Devghat	Koshi, Chatara
Jun	I	1	2	0
	II	2	1	3
Jul	I	11	16	10
	II	13	28	23
Aug	I	38	31	28
	II	43	36	35
Sep	I	8	14	10
	II	6	3	0
Oct	I	6	0	0
	II	0	0	1
Total		128	131	110

for the analysis of the temporal distribution of extreme flows of varying return periods.

4 Results and discussion

4.1 Temporal distribution of the top 1 % extreme flows

Table 2 shows the extreme flows from 1980 to 2015 across months and fortnights. It is evident that the highest frequencies of extreme flows occur between June and September. Specifically, from the second fortnight of June to the first fortnight of September, the majority of the top 1 % extreme flows are recorded, with 90 % (115/128) occurring at Chisapani, 96 % (126/131) at Devghat, and 99 % (109/110) at Chatara. Among the four months, 63 % (81/128), 51 % (67/131), and 57 % (63/110) of the highest 1 % extreme flows occurred in August at Chisapani, Devghat, and Chatara, respectively.

These results reveal several insights. Firstly, the range of magnitudes for the largest 1 % flows at Chisapani, Devghat, and Chatara vary considerably across the stations, which can be attributed to geographical variations and catchment characteristics. In general, the observed concentration of the top 1 % of extreme flows during the monsoon season (June to September) substantiates the widely held notion that the monsoon period is the highest contributor to such events in Nepal. However, a new insight is that the preponderance of extreme flows occurs from the second fortnight of June to the first fortnight of September, which corresponds to the Nepali months of Asadh, Shrawan, and Bhadra rather than the entire 4-month monsoon period. Notably, August emerges as the most pivotal month, accounting for a substantial majority of extreme flood occurrences, possibly due to saturated soil conditions because of sustained heavy rainfall. Given the country's limited infrastructure and technological resources for flood management, our analysis underscores the urgent need for a more dynamic disaster preparedness strategy. Rather than adopting a blanket approach, our findings endorse the allocation of fewer resources in June, increasing in July, peaking in August, and decreasing in September to align with extreme flow patterns.

Table 3. Trend analysis of the annual maxima at the 95 % confidence level (|Zcritical| = 1.96).

Station	MK *Z* value	Sen's slope	Significance status
Chisapani	1.28	53.6	Not significant
Devghat	−0.27	−6.6	Not significant
Chatara	−1.80	−42.6	Not significant

4.2 Temporal distribution of the five highest flows

Figure 2 depicts the five highest flows from each of the 36 years. Results reveal that July and August account for more than 83 %, 81 %, and 84 % of the top five flows at Chisapani, Devghat, and Chatara, respectively.

Table 4 lists the frequency of occurrences of the five highest flows from 1980 to 2015. Furthermore, it is evident that July and August consistently account for a significant majority of these extremes.

Similar to the previous analysis, this examination reinforces the need for a dynamic approach to disaster response and preparedness, ensuring that resources and strategies are strategically allocated to coincide with the heightened flood risk during crucial periods.

4.3 Trend analysis of flood peaks

Trend analysis of the annual maximum of daily flows at a 95 % confidence level showed statistically insignificant trends for all the stations (Table 3). Thus, assuming stationarity in the series, Gumbel's distribution (Chow et al., 1988), being simpler and more popular (Adhikari et al., 2023; Aryal et al., 2020; Dhakal et al., 2023; Dingle et al., 2020), was used for flood frequency analysis in this study.

4.4 Temporal distribution of extreme flows of different return periods

Table 5 lists the distribution of extreme flows by months and fortnights equalling or exceeding floods of different return periods. In the case of Chisapani, a significant proportion of extreme floods, approximately 60 % (28/47), is concentrated in the months of July and August. Notably, August stands as the primary contributor, accounting for 53 % (25/47) of these extremes. Moreover, it is evident that the second fortnight contributes more (52 %, 12/25) to the August floods than the first fortnight (48 %, 13/25). Interestingly, floods

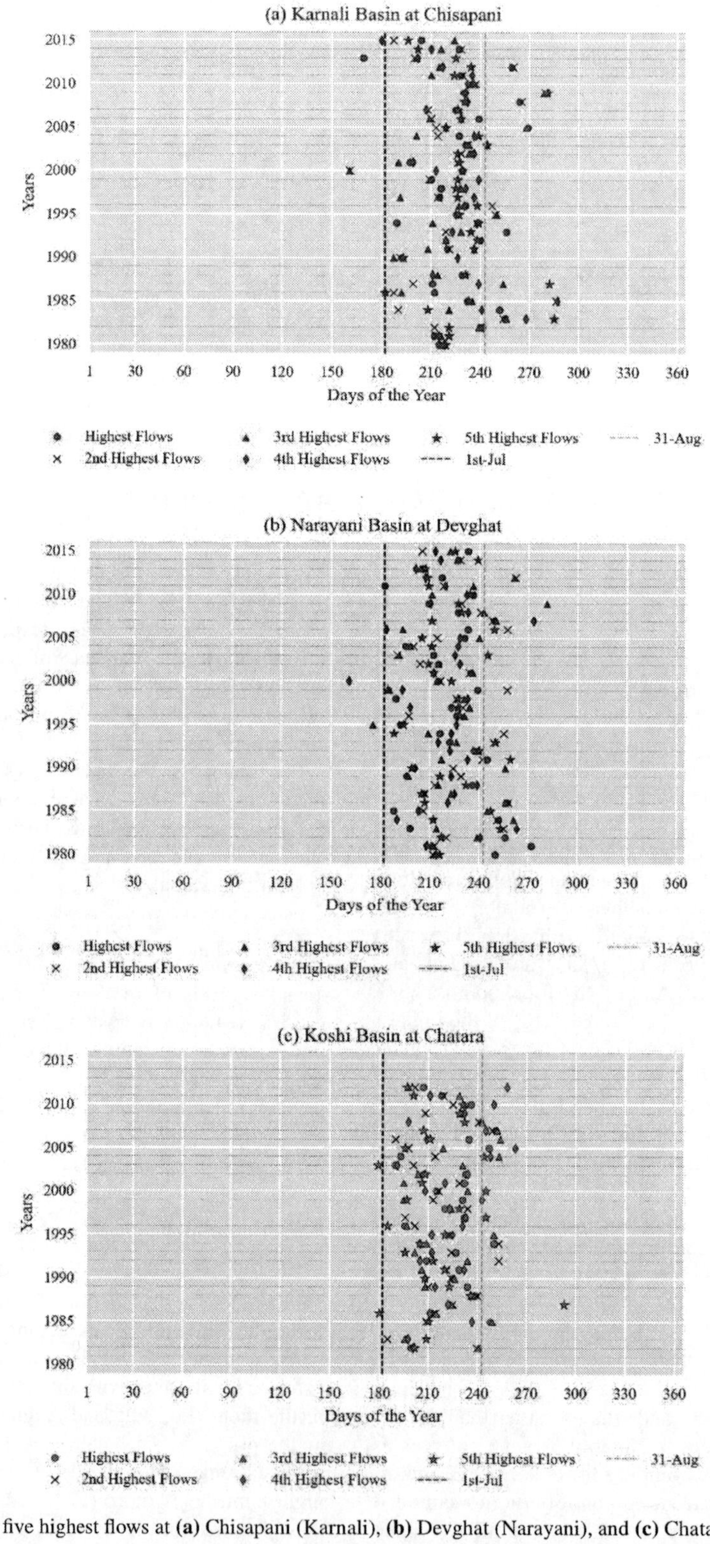

Figure 2. Scatterplot of the five highest flows at **(a)** Chisapani (Karnali), **(b)** Devghat (Narayani), and **(c)** Chatara (Koshi).

Table 4. Occurrences of the five highest flows in July and August.

Station	Month and fortnight	July and August	July I	July II	August I	August II
Chisapani	Five highest flows (total count 175)	146 (83 %)	14 (8 %)	26 (15 %)	48 (27 %)	58 (33 %)
Devghat	Five highest flows (total count 180)	147 (82 %)	17 (9 %)	41 (23 %)	49 (27 %)	40 (22 %)
Chatara	Five highest flows (total count 150)	127 (85 %)	13 (9 %)	44 (29 %)	31 (21 %)	39 (26 %)

Table 5. Temporal distribution of flows exceeding floods of different return periods.

Month	Fortnight	Karnali, Chisapani					Narayani, Devghat					Koshi, Chatara				
		2	5	10	20	100	2	5	10	20	100	2	5	10	20	100
June	I	1	0	0	0	0	0	0	0	0	0	0	0	0	0	0
	II	1	1	1	1	0	0	0	0	0	0	1	0	0	0	0
July	I	2	0	0	0	0	7	0	0	0	0	8	2	1	1	0
	II	1	0	0	0	0	7	4	0	0	0	18	6	1	0	0
August	I	7	2	1	1	1	8	1	0	0	0	15	3	1	1	1
	II	10	2	1	0	0	10	4	2	1	1	25	4	1	0	0
September	I	3	3	2	1	0	4	1	0	0	0	4	1	1	0	0
	II	1	0	0	0	0	1	0	0	0	0	0	0	0	0	0
October	I	2	1	1	0	0	0	0	0	0	0	0	0	0	0	0
	II	0	0	0	0	0	0	0	0	0	0	1	0	0	0	0
Total		28	9	6	3	1	37	10	2	1	1	72	16	5	2	1

with higher return periods (20- and 100-year) tend to occur more frequently in the first fortnight of August. Similarly, for Devghat, the majority of extreme floods, approximately 88 % (45/51), are concentrated in the months of July and August. August again contributes most (53 %, 27/51) of these extreme events. Similarly to Chisapani, the second fortnight accounts for 67 % (18/27) of August's extreme floods, compared to the first fortnight (33 %, 9/27).

It is notable that the occurrence of floods with higher return periods (20- and 100-year) is more frequent in the second fortnight of August. In the case of Chatara, a remarkable 92 % (88/96) of the extreme floods occur in July and August combined, with August accounting for 53 % (51/96) of these extreme events. A closer examination reveals that the second fortnight has a larger contribution of 59 % (30/51) of the extreme floods. Similar to the case of Chisapani, floods of 20 and 100 years tend to occur more frequently in the first fortnight of August.

These results reveal interesting flood patterns across the three stations. As anticipated, the magnitude of peak floods escalates, while the frequency is reduced with larger return periods. Nevertheless, the most remarkable revelation is the prominence of August, which emerges as the month exhibiting the most concentrated instances of extreme floods. These findings have significant implications for disaster response and preparedness in Nepal, underscoring a greater focus on the month of August.

5 Conclusions

This study analyzed the temporal distribution of the top 1 % of extreme flows and the five highest flows annually, and it examined the occurrences of extreme flows of different return periods of Nepal at fortnightly resolution.

Our results demonstrate substantial variability in flood magnitudes across Chisapani, Devghat, and Chatara, with a pronounced concentration of extreme flows during the second fortnight of June to the first fortnight of September, with August being the most critical month. Additionally, the temporal trends shown by the flood peaks are not significant. Moreover, the temporal distribution of floods of different return periods emphasizes the pivotal significance of August for effective flood management. We conclude that a dynamic approach to flood disaster preparedness and response is required for Nepal. Under conditions of constrained resources, focus on strategic resource allocation that peaks in August is preferable for enhancing disaster resilience and optimizing flood management efforts. Future studies could benefit from expanding the analysis to include an extended number of gauging stations across the river network for devising localized disaster management strategies.

Data availability. The flow data were obtained from the Department of Hydrology and Meteorology, Government of Nepal (https://www.dhm.gov.np/, last access: 31 July 2024).

Author contributions. AB did the conceptualization, did the major analysis, and wrote the first draft of the manuscript. UB reviewed the manuscript. KRM supported the analysis. LPD reviewed the final manuscript.

Competing interests. The contact author has declared that none of the authors has any competing interests.

Disclaimer. Publisher's note: Copernicus Publications remains neutral with regard to jurisdictional claims made in the text, published maps, institutional affiliations, or any other geographical representation in this paper. While Copernicus Publications makes every effort to include appropriate place names, the final responsibility lies with the authors.

Special issue statement. This article is part of the special issue "Mountain Hydrology and Cryosphere". It is a result of the International Conference on Mountain Hydrology and Cryosphere, Kathmandu and Dhulikhel, Nepal, 9–10 November 2023.

References

Adhikari, T. R., Baniya, B., Tang, Q., Talchabhadel, R., Gouli, M. R., Budhathoki, B. R., and Awasthi, R. P.: Evaluation of post extreme floods in high mountain region: A case study of the Melamchi flood 2021 at the Koshi River Basin in Nepal, Nat. Hazards Res., 3, 437–446, https://doi.org/10.1016/j.nhres.2023.07.001, 2023.

Aryal, D., Wang, L., Adhikari, T. R., Zhou, J., Li, X., Shrestha, M., Wang, Y., and Chen, D.: A model-based flood hazard mapping on the southern slope of Himalaya, Water (Switzerland), 12, 540, https://doi.org/10.3390/w12020540, 2020.

Bhattarai, R., Bhattarai, U., Pandey, V. P., and Bhattarai, P. K.: An artificial neural network-hydrodynamic coupled modeling approach to assess the impacts of floods under changing climate in the East Rapti Watershed, Nepal, J. Flood Risk Manag., 15, e12852, https://doi.org/10.1111/jfr3.12852, 2022.

Chow, V. Te, Maidment, D. R., and Mays, L. W.: Applied hydrology, McGraw-Hill, 572 pp., ISBN 0070108102, 1988.

Devkota, P., Dhakal, S., Shrestha, S., and Shrestha, U. B.: Land use land cover changes in the major cities of Nepal from 1990 to 2020, Environ. Sustain. Indic., 17, 100227, https://doi.org/10.1016/j.indic.2023.100227, 2023.

Devkota, R., Bhattarai, U., Devkota, L., and Maraseni, T. N.: Assessing the past and adapting to future floods: a hydro-social analysis, Clim. Change, 163, 1065–1082, https://doi.org/10.1007/s10584-020-02909-w, 2020.

Devkota, R. P. D.: Flood Vulnerability through the Eyes of Vulnerable People in Mid-Western Terai of Nepal, J. Earth Sci. Clim. Change, 4, 132, https://doi.org/10.4172/2157-7617.1000132, 2013.

Dhakal, S., Bhattarai, U., Marahatta, S., and Devkota, P.: Impact of climate change on the full spectrum of future low flows of Budhigandaki River Basin in Nepal using Gumbel distribution, Int. J. Energy Water Resour., 7, 191–203, https://doi.org/10.1007/s42108-022-00214-z, 2023.

Dingle, E. H., Creed, M. J., Sinclair, H. D., Gautam, D., Gourmelen, N., Borthwick, A. G. L., and Attal, M.: Dynamic flood topographies in the Terai region of Nepal, Earth Surf. Proc. Landf., 45, 3092–3102, https://doi.org/10.1002/esp.4953, 2020.

Gautam, D. K. and Dulal, K.: Determination of Threshold Runoff for Flood Warning in Nepalese Rivers, Journal of Integrated Disaster Risk Management, 3, 125–136, https://doi.org/10.5595/idrim.2013.0061, 2013.

Khatiwada, K. R., Panthi, J., Shrestha, M. L., and Nepal, S.: Hydro-climatic variability in the Karnali River Basin of Nepal Himalaya, Climate, 6, 74, https://doi.org/10.3390/cli4020017, 2018.

Luo, T., Maddocks, A., Iceland, C., Ward, P., and Winsemius, H.: World's 15 Countries with the Most People Exposed to River Floods, https://www.wri.org/insights/worlds-15-countries-most-people-exposed-river-floods (last access: 30 July 2024), 2015.

Mann, H. B.: Nonparametric Tests Against Trend, Econometrica, 13, 245–259, 1945.

National Planning Commission: Nepal Flood 2017: Post Flood Recovery Needs Assessment, Government of Nepal, Kathmandu, https://un.org.np/sites/default/files/doc_publication/2018-11/PFRNA_Report_Final.pdf (last access: 30 July 2024), 2017.

World Meteorological Organization: WMO atlas of mortality and economic losses from weather, climate and water extremes, 90 pp., ISBN 9789263112675, 2019.

Proc. IAHS, 387, 9–15, 2024
https://doi.org/10.5194/piahs-387-9-2024

Impact of changes in climate and glacier configurations on runoff from the Langtang River basin, Nepal

Dhiraj Pradhananga[1,2], Susa Manandhar[1,2], Bharat Dhungana[1,2], Manavi Chaulagain[1,2], Bhola Nath Dhakal[2,3], and Sunil Adhikary[1]

[1]Department of Meteorology, Tri-Chandra Multiple Campus, Tribhuvan University, Kathmandu, Nepal
[2]The Small Earth Nepal, Kathmandu, Nepal
[3]Department of Geography, Ratna Rajya Laxmi Campus, Tribhuvan University, Kathmandu, Nepal

Correspondence: Dhiraj Pradhananga (dhiraj.pradhananga@trc.tu.edu.np)

Published: 18 November 2024

Abstract. Climate change is rapidly altering Himalayan glaciers, jeopardizing downstream water resources. This study investigates the impact of changing air temperatures, precipitation, and glacier configurations on streamflow in the Langtang River basin, a vital Himalayan basin. Using a process-based glacio-hydrological model driven by bias-corrected reanalysis data, we simulate streamflow under various scenarios. Results reveal that, while current glacier conditions show increased runoff with higher temperatures and precipitation, future deglaciation scenarios project a decrease in total streamflow, highlighting the complex interplay between climate and glacier dynamics in basin hydrology. These findings emphasize the need for adaptation strategies to ensure water security in the face of evolving Himalayan hydrology.

1 Introduction

Mountain snow and glaciers, nature's frozen reservoirs, play a vital role in maintaining downstream freshwater resources. The storage and release of water depend on climatological factors, making them sensitive indicators of climate change. Rapid changes in climatic conditions have shifted the accumulation and melting patterns of snow field areas and glaciers. Their retreat is impacting water resources across the globe as snow and ice have a strong influence on their basin hydrological regime due to their capacity to store water on seasonal to decadal timescales (Jansson et al., 2003).

Runoff dynamics in glacierized and non-glacierized catchments are very different, and it is challenging to understand the interconnection between meteorological parameters and to incorporate them into the model. Mountain glacierized headwaters are undergoing a transient shift in their hydrological and glaciological systems due to rapid climate change. In this research, we set out to understand how the hydrology and streamflow generation in glacierized basins will be affected due to shifts in air temperature and precipitation patterns.

In glacierized areas, snow and exposed or debris-covered ice play a major role in streamflow generation, but these factors are often neglected in hydrological models (Koboltschnig and Schöner, 2011; van Tiel et al., 2020). For a glacierized basin like Langtang, without the proper modeling approaches, it is very difficult to understand the impact of climate change and the contribution of glaciers to the runoff. While looking at the changes and shifts in glaciers, only meteorological and climatic factors are considered. Physical factors such as land cover are the least considered when estimating the influence of glacier melt in glacier hydrology (Aubry-Wake et al., 2022).

By applying a process-oriented glacial–hydrological model, this study quantifies the evolving influences of on-glacier and off-glacier processes on streamflow generation in the context of future climate scenarios. This approach aims to safeguard freshwater availability from glacierized basins for the well-being of dependent ecosystems and communities.

Published by Copernicus Publications on behalf of the International Association of Hydrological Sciences.

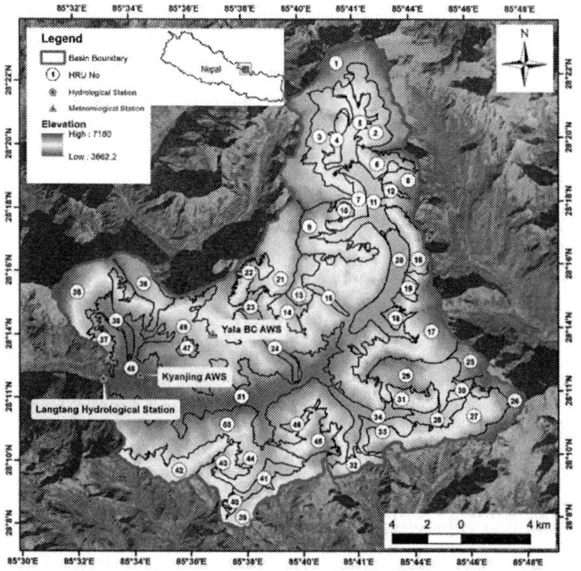

Figure 1. Location map of Langtang River basin, Rasuwa District, Nepal. A map of Nepal is shown in the inset.

2 Materials and methods

2.1 Study area

The Langtang River basin (LRB) in the central Himalayas of Nepal plays a vital role in sustaining downstream communities and ecosystems. The LRB lies approximately 100 km north of Kathmandu, bordering between Nepal and China (Fig. 1). It spans from 28°08′ to 28°23′ N latitude and from 85°35′ to 85°48′ E longitude, covering an area of 354 km^2. The elevation ranges from 3800 m a.s.l. up to the peak of Langtang Lirung at 7234 m a.s.l, with an average altitude of 5168 m a.s.l, a mean slope of 26.7°, and a predominantly south-facing aspect (mean 193.5°). In 2000, the area of the glacier was 189 km^2 (53 %), and the accumulation area ratio (AAR) was 0.8. The main valley, carved by glacial activity into a characteristic U shape, is divided by the Langtang River (Adhikari et al., 2014). Rising concerns about climate change and its potential impact on the LRB's glaciers and water resources motivated this study to investigate the complex interplay between climate, glaciers, and streamflow generation.

2.2 Meteorological and hydrological data

This study utilized both observed and reanalysis meteorological data to drive the Cold Regions Hydrological Model (CRHM). Daily and hourly data from Kyangjing Station (Fig. 2), spanning the periods 2002–2009 and 2012–2019, provide essential information on temperature, precipitation, solar radiation, relative humidity, and wind speed, obtained from the Department of Hydrology and Meteorology (DHM) and the International Centre for Integrated Mountain Development (ICIMOD). Due to limitations in the availability of observed data, we employed bias-corrected reanalysis data from the Water and Global Change (WATCH) forcing data methodology applied to ERA-Interim data (WFDEI, Weedon et al., 2011). WFDEI was chosen for its superior representation of the LRB's climatic conditions compared to other reanalysis datasets (ERA-Interim and MERRA-2), particularly in terms of accurately capturing precipitation patterns and temperature fluctuations. We validated the CRHM model with observed streamflow at Langtang Khola (Fig. 2b).

Spatial data such as land cover and soil maps were obtained from the Topographical Survey and Land Use Management Division, as well as from the ASTER Global Digital Elevation Model (DEM), with a resolution of 30 m; these were used to characterize the basin's physical characteristics and to provide inputs for the CRHM Model.

2.3 Cold Regions Hydrological Model (CRHM)

This study employs the CRHM platform (Pomeroy et al., 2022), a flexible framework for developing process-based hydrological models tailored to cold regions and temperate climates. Its modular structure allowed us to incorporate CRHM-Glacier (Pradhananga and Pomeroy, 2022a), a specific model extension equipped with modules for simulating key glacier-related processes like firnification, snow redistribution by wind and gravity, and coupled mass and energy budgets. These features are crucial for accurately representing the complex runoff dynamics of the LRB, influenced by both snowmelt and glacier melt. Figure 3 illustrates the modular structure of CRHM-Glacier, highlighting the specific modules used in this study.

2.4 Methodology

To investigate the impact of climate change and glacier dynamics on LRB streamflow, we adopted a multi-pronged approach. The CRHM-Glacier model built for LRB was considered for the model runs with different possible scenarios. We divided the basin into 51 hydrological response units (HRUs) to capture spatial variations in topography, land cover, and glacier distribution; details are in Pradhananga and Pomeroy (2022a). The bias-corrected WFDEI data of air temperature, humidity, precipitation, wind speed, and incoming shortwave and longwave radiation were used to force the CRHM-glacier. We simulated streamflow from 2008 to 2018 and presented the aggregated values. We conducted sensitivity analyses to assess the model's response to changes in individual parameters. For example, we increased and decreased precipitation values to understand the model's sensitivity to future precipitation variations. We also ran simulations under different future climate scenarios and glacier configurations (similar to Pradhananga and Pomeroy, 2022b) to predict po-

Figure 2. Photos showing **(a)** Kyangjing meteorological station and **(b)** hydrometric station at Langtang River.

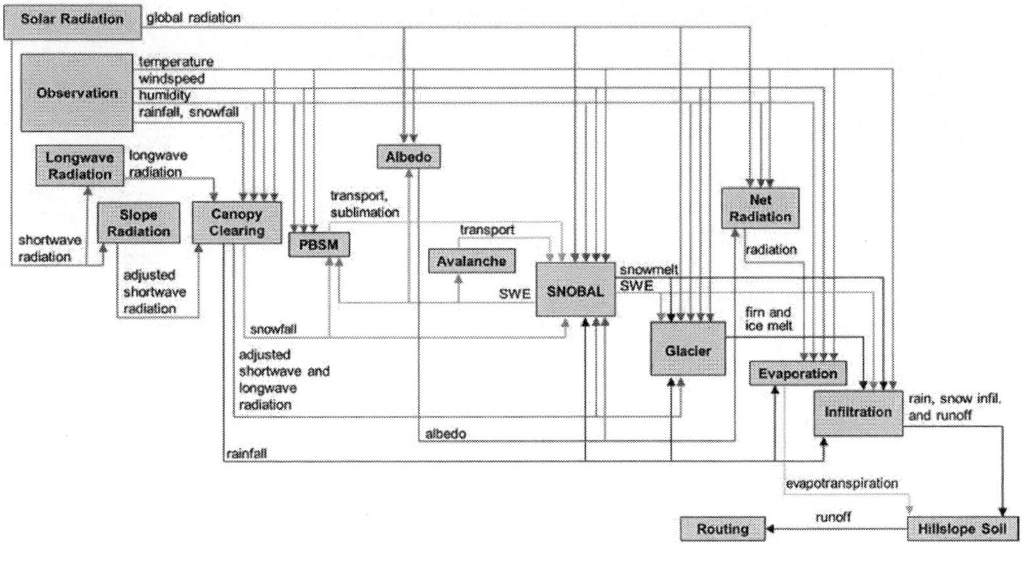

Figure 3. Modular structure of CRHM-glacier (Pradhananga and Pomeroy, 2022a).

tential changes in runoff patterns under varying temperature and precipitation regimes.

3 Results and discussion

3.1 Glacier configuration

This study employed the CRHM-Glacier to investigate the interplay between glacier configuration and streamflow in the LRB under both present and future climate scenarios. Our key findings highlight the complex changes in runoff patterns resulting from glacier retreat and climatic shifts.

3.1.1 Present climatic scenario

Figure 4a visualizes the simulated daily averaged streamflow for the period 2008–2018 with and without glaciers, representing the LRB's current and deglaciated state. Notably, removing glaciers in the simulation resulted in a 32 % decrease in total annual runoff (Fig. 4b). This finding contradicts the intuitive assumption that more glacier melt would lead to higher runoff. However, it can be attributed to the crucial role glacial ice plays in storing and releasing water over the course of a year. During the warm seasons, meltwater from glaciers supplements streamflow, compensating for reduced snowmelt and mitigating summer droughts. This underscores the vital contribution of glaciers to the LRB's seasonal water balance beyond simply providing additional runoff through melting.

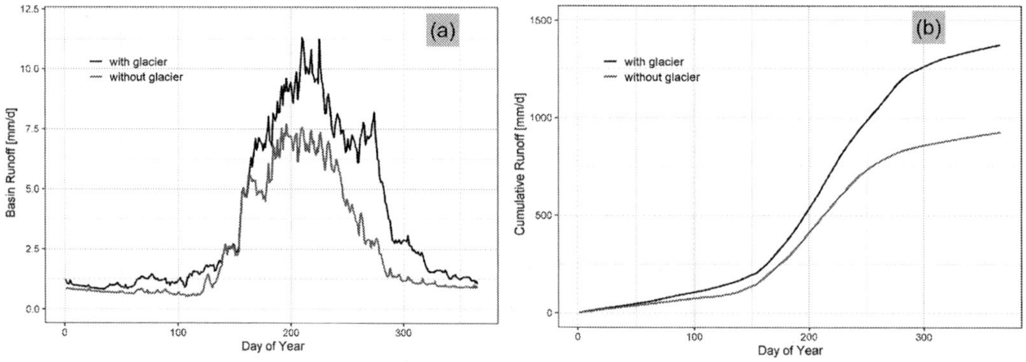

Figure 4. Variation in streamflow with and without glaciers in LRB: **(a)** daily averaged values for the period 2008–2018 and **(b)** yearly cumulative runoff value.

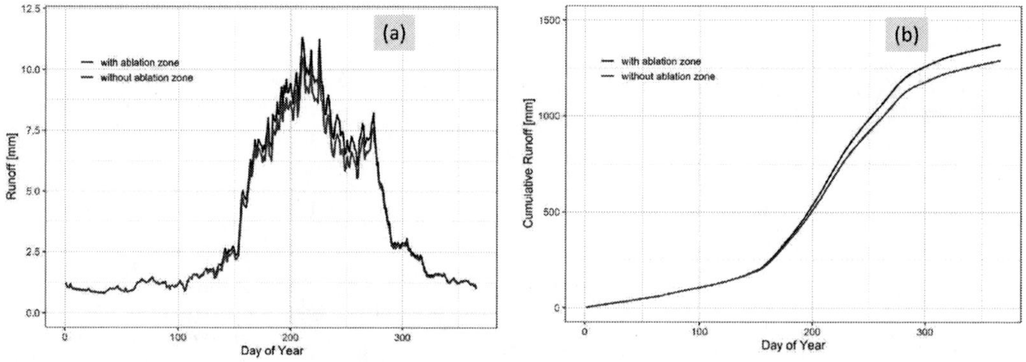

Figure 5. Variation in streamflow with and without ablation zone in LRB: **(a)** daily averaged values for the period 2008–2018 and **(b)** yearly cumulative runoff value.

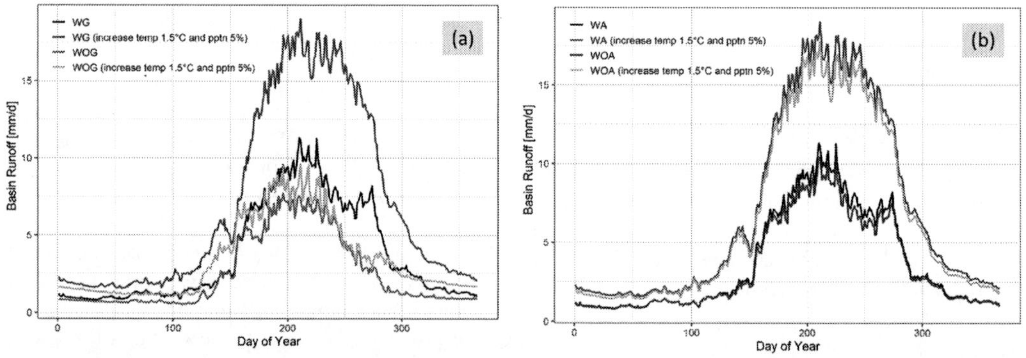

Figure 6. Daily averaged values of streamflow for the period 2008–2018 **(a)** with and without glaciers and **(b)** with and without ablation zone.

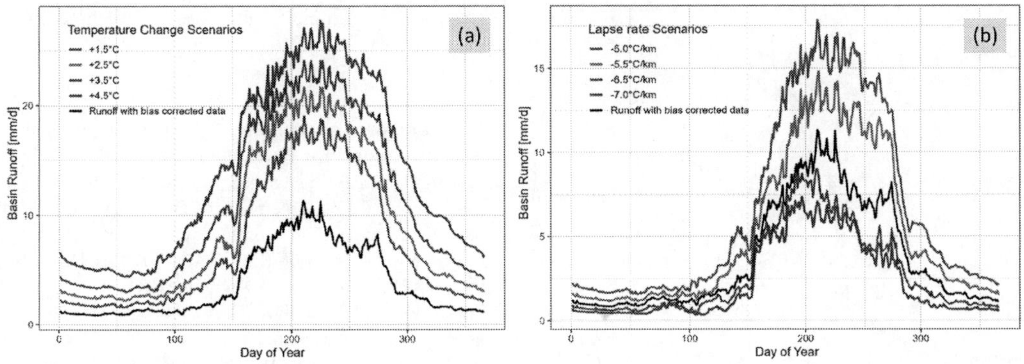

Figure 7. Daily averaged simulation of streamflow for the period 2008–2018. **(a)** Air temperature scenario. **(b)** Lapse rate scenario.

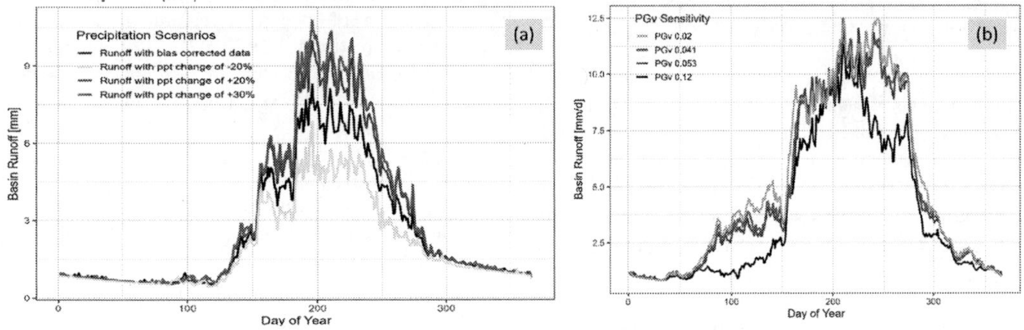

Figure 8. Daily averaged simulation of streamflow for the period 2008–2018. **(a)** Precipitation scenario. **(b)** Precipitation gradient scenario.

Figure 5 further explores the impact of the ablation zone, the lowermost region of a glacier. Simulating scenarios with and without an ablation zone, which comprises 10.7 % of the total basin area (representing potential future retreat), suggests a 7 % decrease in total annual runoff. This highlights the importance of preserving the ablation zone to maintain water resources in the face of climate change.

3.1.2 Future climate scenario

There are high projections that the glacier melt would be around 80 % in the worst-case scenarios (IPCC, 2023; Khadka et al., 2020; Prasad et al., 2019). Hence, we are looking into the condition for stabilizing the temperature by 1.5 °C, which might increase the precipitation by 5 %. In Fig. 6a, considering the climatic scenario for the extreme conditions with and without glaciers, there is a decrease in the runoff by −7.9 %. When taking the condition of Fig. 6b, which is, in this case, just the ablation-zone retreat, there would be an increase in runoff by 77 %. The model does show the trend of high runoff during the monsoon and a decrease in runoff to below normal during the dry period, providing a rough estimate that, in the long run, with increasing trends in climate change, the wet period will be wetter, and the dry period will be dryer.

3.2 Sensitivity analysis

To delve deeper into the model's response to individual parameters, we conducted sensitivity analyses for air temperature, lapse rate, precipitation, precipitation gradient, and firn albedo (Figs. 7–9). Notably, the model exhibited the highest sensitivity towards air temperature, with increasing temperature leading to a proportional increase in runoff. This aligns with established knowledge of glacier melt dynamics and its impact on streamflow. Similarly, a decrease in runoff was simulated with an increasing lapse rate. This suggests that, while warming temperatures might initially increase runoff, their impact can be moderated by atmospheric temperature gradients within the basin.

The other sensitivity analyses revealed a positive correlation between runoff and precipitation, a moderate sensitivity to the precipitation gradient, and minimal changes with varying firn albedo (Fig. 9).

This study highlights the profound influence of glacier configuration and climate change on streamflow patterns in the LRB. Our findings offer critical insights into the com-

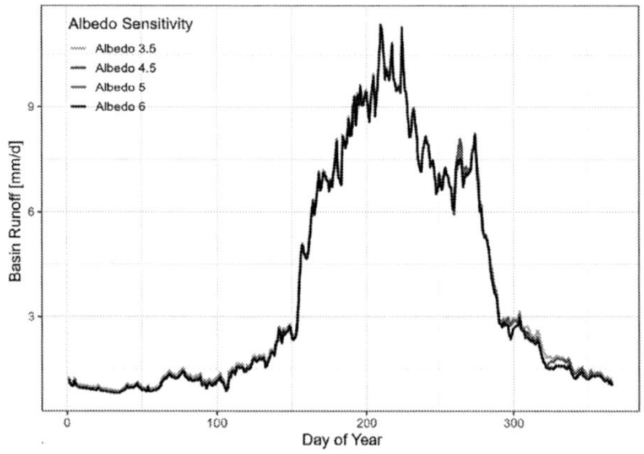

Figure 9. Daily average simulation of streamflow with changes in firn albedo scenarios for the period 2008–2018.

plex interplay between glaciers, temperature, precipitation, and seasonal water availability. The rapid decrease in runoff under complete deglaciation emphasizes the crucial role of glacier storage in maintaining the basin's hydrological balance. Future climate scenarios suggest mixed effects, with potential increases in runoff during wet periods but concerns about further reductions during dry seasons. These findings highlight the urgent need for effective water management strategies that adapt to evolving glacier dynamics and changing climate patterns to ensure water security for downstream communities.

4 Conclusion

This study utilized a process-based glacio-hydrological model, CRHM-Glacier, to simulate streamflow in the Langtang River basin under various climate and glacier change scenarios. The complete removal of glaciers resulted in a 32 % decrease in total runoff, highlighting the critical role of glaciers in maintaining mountain water resources. Future climate scenarios with increased temperature and precipitation also showed substantial alterations in runoff patterns, with dry seasons becoming drier and wet seasons becoming wetter.

This study further confirmed the positive correlation between precipitation and runoff, highlighting their critical role in shaping basin hydrology. Additionally, sensitivity analyses revealed the influence of the lapse rate, precipitation gradient, and firn albedo on streamflow dynamics. These findings emphasize the urgency of understanding and adapting to the evolving hydrological regime of the LRB for the long-term sustainability of water resources in the region. Further research is essential to understand how glacier dynamics, in combination with regional climate patterns and land-use changes, impact runoff patterns. This understanding is critical for developing informed long-term water management strategies in vulnerable mountainous regions like the Langtang Valley.

Code and data availability. The hydrological and meteorological data used for this study are available from the Regional Database System of ICIMOD at https://rds.icimod.org/ (last access: 13 September 2024). The meteorological data for the Kyanging Meteorological Station can be accessed at https://doi.org/10.26066/RDS.22464 (ICIMOD, 2016), and the hydrological data for Langtang River can be accessed at https://doi.org/10.26066/RDS.1972407 (ICIMOD and Utrecht University, 2020). The reanalysis data (WFDEI) can be downloaded from https://doi.org/10.5065/486N-8109 (Weedon et al., 2018). The CRHM software source code and its installation instructions are available on the CHRM GitHub page at https://github.com/srlabUsask/crhmcode/releases/tag/release_v1.3.5 (last access: 13 September 2024, Centre for Hydrology, 2024a). Further details on the model platform and related publications are available at https://research-groups.usask.ca/hydrology/modelling/crhm.php (last access: 13 September 2024, Centre for Hydrology, 2024b).

Author contributions. DP conceptualized the main idea and the CRHM project for LRB, prepared the meteorological forcing data, and led the preparation of the paper. SM, BD, and MC developed and generated the model simulations, analyzed the results, and contributed to drafting the paper. BND and SA contributed to data collection and reviewed the paper.

Competing interests. At least one of the (co-)authors is a guest member of the editorial board of *Proceedings of IAHS* for the special issue "Mountain Hydrology and Cryosphere". The peer-review process was guided by an independent editor, and the authors also have no other competing interests to declare.

Disclaimer. Publisher's note: Copernicus Publications remains neutral with regard to jurisdictional claims made in the text, published maps, institutional affiliations, or any other geographical representation in this paper. While Copernicus Publications makes every effort to include appropriate place names, the final responsibility lies with the authors.

Special issue statement. This article is part of the special issue "Mountain Hydrology and Cryosphere". It is a result of the International Conference on Mountain Hydrology and Cryosphere, Kathmandu and Dhulikhel, Nepal, 9–10 November 2023.

Acknowledgements. We are deeply grateful to ICIMOD for providing access to their open-source hydro-meteorological data, which played a crucial role in our research. We extend our sincere thanks to John Pomeroy and Xing Fang of the University of Saskatchewan and Caroline Aubry-Wake of the University of Utrecht for their invaluable support in applying the CRHM model to the Langtang River basin in Nepal. We acknowledge the dedicated efforts of Nammy Hang Kirat, Piyush Dahal, Kajol Basnet, Nirab Shrestha, and Sujan Khanal of The Small Earth Nepal (SEN) for their assistance with GIS map production and data collection and analysis.

Financial support. This research has been supported by the University Grants Commission of Nepal (grant no. CRG-76/77-S&T-1) and the Nepal Academy of Science and Technology (NAST) research grant 2018 to Dhiraj Pradhananga.

References

Adhikari, T. R., Devkota, L. P., and Shrestha, A. B.: Climate change scenarios and its impact on water resources of Langtang Khola Basin, Nepal, Proc. IAHS, 364, 9–13, https://doi.org/10.5194/piahs-364-9-2014, 2014.

Aubry-Wake, C., Pradhananga, D., and Pomeroy, J. W.: Hydrological process controls on streamflow variability in a glacierized headwater basin, Hydrol. Process., 36, e14731, https://doi.org/10.1002/hyp.1473, 2022.

Centre for Hydrology: CRHM: Cold Regions Hydrological Model (Version 1.3.5), University of Saskatchewan [code], https://github.com/srlabUsask/crhmcode/releases/tag/release_v1.3.5 (last access: 13 September 2024), 2024a.

Centre for Hydrology: Modelling CRHM: The Cold Regions Hydrological Model, University of Saskatchewan [code], https://research-groups.usask.ca/hydrology/modelling/crhm.php (last access: 13 September 2024), 2024b.

ICIMOD: Meteorological data from Kyanging automatic weather station, ICIMOD [data set], https://doi.org/10.26066/RDS.22464, 2016.

ICIMOD and Utrecht University: Discharge: Langtang Khola, ICIMOD [data set], https://doi.org/10.26066/RDS.1972407, 2020.

Intergovernmental Panel on Climate Change (IPCC): Climate Change 2021 – The Physical Science Basis: Working Group I Contribution to the Sixth Assessment Report of the Intergovernmental Panel on Climate Change, 1211–1362, Cambridge Univ. Press, https://doi.org/10.1017/9781009157896.011, 2023.

Jansson, P., Hock, R., and Schneider, T.: The concept of glacier storage: A review, J. Hydrol., 282, 116–129, https://doi.org/10.1016/S0022-1694(03)00258-0, 2003.

Khadka, M., Kayastha, R. B., and Kayastha, R.: Future projection of cryospheric and hydrologic regimes in Koshi River basin, Central Himalaya, using coupled glacier dynamics and glacio-hydrological models, J. Glaciol., 66, 831–845, https://doi.org/10.1017/jog.2020.51, 2020.

Koboltschnig, G. R. and Schöner, W.: The relevance of glacier melt in the water cycle of the Alps: the example of Austria, Hydrol. Earth Syst. Sci., 15, 2039–2048, https://doi.org/10.5194/hess-15-2039-2011, 2011.

Pomeroy, J. W., Brown, T., Fang, X., Shook, K. R., Pradhananga, D., Armstrong, R., Harder, P., Marsh, C., Costa, D., Krogh, S. A., Aubry-Wake, C., Annand, H., Lawford, P., He, Z., Kompanizare, M., and Lopez Moreno, J. I.: The cold regions hydrological modelling platform for hydrological diagnosis and prediction based on process understanding, J. Hydrol., 615, 128711, https://doi.org/10.1016/j.jhydrol.2022.128711, 2022.

Pradhananga, D. and Pomeroy, J. W.: Diagnosing changes in glacier hydrology from physical principles using a hydrological model with snow redistribution, sublimation, firnification and energy balance ablation algorithms, J. Hydrol., 608, 127545, https://doi.org/10.1016/j.jhydrol.2022.127545, 2022a.

Pradhananga, D. and Pomeroy, J. W.: Recent hydrological response of glaciers in the Canadian Rockies to changing climate and glacier configuration, Hydrol. Earth Syst. Sci., 26, 2605–2616, https://doi.org/10.5194/hess-26-2605-2022, 2022b.

Prasad, V., Kulkarni, A. V., Pradeep, S., Pratibha, S., Tawde, S. A., Shirsat, T., Arya, A. R., Orr, A., and Bannister, D.: Large losses in glacier area and water availability by the end of twenty-first century under high emission scenario, Satluj basin, Himalaya. Curr. Sci., 116, 1721–1730, https://doi.org/10.18520/cs/v116/i10/1721-1730, 2019.

van Tiel, M., Stahl, K., Freudiger, D., and Seibert, J.: Glacio-hydrological model calibration and evaluation, WIREs Water, 7, e1483, https://doi.org/10.1002/wat2.1483, 2020.

Weedon, G. P., Gomes, S., Viterbo, P., Shuttleworth, W. J., Blyth, E., Österle, H., Adam, J. C., Bellouin, N., Boucher, O., and Best, M.: Creation of the WATCH Forcing Data and Its Use to Assess Global and Regional Reference Crop Evaporation over Land during the Twentieth Century, J. Hydrometeorol., 12, 823–848, https://doi.org/10.1175/2011JHM1369.1, 2011.

Weedon, G. P., Balsamo, G., Bellouin, N., Gomes, S., Best, M. J., and Viterbo, P.: The WFDEI Meteorological Forcing Data, Research Data Archive at the National Center for Atmospheric Research, Computational and Information Systems Laboratory [data set], https://doi.org/10.5065/486N-8109, 2018.

Proc. IAHS, 387, 17–24, 2024
https://doi.org/10.5194/piahs-387-17-2024

A hybrid approach to enhance streamflow simulation in data-constrained Himalayan basins: combining the Glacio-hydrological Degree-day Model and recurrent neural networks

Dinesh Joshi[1], **Rijan Bhakta Kayastha**[1], **Kundan Lal Shrestha**[2], **and Rakesh Kayastha**[1]
[1]Department of Environmental Science and Engineering, School of Science, Kathmandu University, Dhulikhel, 45210, Nepal
[2]Department of Chemical Science and Engineering, School of Engineering, Kathmandu University, Dhulikhel, 45210, Nepal
Correspondence: Dinesh Joshi (joshidinesh0227@gmail.com)

Published: 18 November 2024

Abstract. The Glacio-hydrological Degree-day Model (GDM) is a distributed model, but it is prone to uncertainties due to its conceptual nature, parameter estimation, and limited data in the Himalayan basins. To enhance accuracy without sacrificing interpretability, we propose a hybrid model approach that combines GDM with recurrent neural networks (RNNs), hereafter referred to as GDM–RNN. Three RNN types – a simple RNN model, a gated recurrent unit (GRU) model, and a long short-term memory (LSTM) model – are integrated with GDM. Rather than directly predicting streamflow, RNNs forecast GDM's residual errors. We assessed performance across different data availability scenarios, with promising results. Under limited-data conditions (1 year of data), GDM–RNN models (GDM–simple RNN, GDM–LSTM, and GDM–GRU) outperformed standalone GDM and machine learning models. Compared with GDM's respective Nash–Sutcliffe efficiency (NSE), R^2, and percent bias (PBIAS) values of 0.80, 0.63, and −4.78, the corresponding values for the GDM–simple RNN were 0.85, 0.82, and −6.21; for GDM–LSTM, they were 0.86, 0.79, and −6.37; and for GDM–GRU, they were 0.85, 0.8, and −5.64. Machine learning models yielded similar results, with the simple RNN at 0.81, 0.7, and −16.6; LSTM at 0.79, 0.65, and −21.42; and GRU at 0.82, 0.75, and −12.29, respectively. Our study highlights the potential of machine learning with respect to enhancing streamflow predictions in data-scarce Himalayan basins while preserving physical streamflow mechanisms.

1 Introduction

Hydrological models, classified as spatially distributed or lumped, differ in complexity and scope. While spatially distributed models cover diverse parameters across an area, lumped models use simpler relationships with fewer variables (Refsgaard and Knudsen, 1996). Physically based models aim for accuracy but may struggle with fine-scale features, often operating at larger scales with indirect parameter estimation (Beven, 2002). Despite their grounding in physical principles, they may not precisely represent all hydrological processes (Beven, 2002).

In high-mountain regions, like the Himalayas, distributed rainfall–runoff models face challenges due to the inadequate representation of snow and glacial processes. Glacio-hydrological models are crucial, but their effectiveness relies on data accessibility, with results influenced by the spatial and temporal resolutions (Réveillet et al., 2018). Despite the notion that increased data accessibility simplifies model construction, it often reveals the intricate nature of hydrological systems across diverse temporal and spatial scales (De Filippis et al., 2020).

Addressing inverse problems for calibrated hydrological models faces challenges like nonuniqueness and computa-

Published by Copernicus Publications on behalf of the International Association of Hydrological Sciences.

tional demands (Carrera et al., 2005). Uncertainties in hydrological models arise from observation accuracy, model variations, parameter distributions, estimation, and forcing variables (Dahlstrom, 2015).

Machine learning algorithms aid parameter estimation, especially with abundant data (Solomatine, 2006). While physically derived models persist for comprehensive insights (Nearing et al., 2020), machine learning models complement them by addressing unpredictability or uncertainty (Solomatine, 2006). However, machine learning models lack interpretability as black-box models and may struggle in data-scarce regions like the Himalayan basin (Kayastha and Kayastha, 2019; Jia et al., 2020).

Debates on river discharge modeling focus on understanding physical processes vs. replicating observed values (Ji et al., 2021). An emerging solution combines both methods, using theory-based elements alongside data science to model residual errors. This approach constructs a residual error model to represent the theory-based model's discrepancies.

This research investigates the efficacy of recurrent neural networks (RNNs) with respect to modeling hydrological residual errors for improved streamflow predictions in data-constrained Himalayan basins. Specifically, the performance of different RNNs – simple RNN, a gated recurrent unit (GRU) model, and a long short-term memory (LSTM) model – is compared in contrast to a previous study's focus on LSTM alone (Cho and Kim, 2022). Simulations using the Glacio-hydrological Degree-day Model (GDM), simple RNN, GRU, and LSTM individually are also compared with proposed models (GDM–simple RNN, GDM–GRU, and GDM–LSTM).

2 Material and methods

2.1 Study area

The study area, Marsyangdi River basin in the Nepal Himalayas (Fig. 1), spans 4059 km^2, ranging from 355 to 7819 m a.s.l. (above sea level). About 13.3 % of the area is glacier-covered terrain, mainly between 4000 and 6500 m a.s.l. The basin experiences the Indian summer monsoon (June–September) and occasional westerly disturbances post-monsoon (October–January). Geographically, it features diverse terrain, primarily on the southern slopes of the Central Himalayas, influenced by the Annapurna Massif in the northwest.

2.2 Glacio-hydrological Degree-day Model (GDM)

The Glacio-hydrological Degree-day Model (GDM, version 1.0) is a gridded hydrological model that evaluates the impact of various hydrological elements on river discharge (Kayastha and Kayastha, 2019). Operating daily, it incorporates snowmelt, glacier ice melt, rainfall, and baseflow runoff components. GDM utilizes a degree-day method for glacier ice melt and snowmelt, simplifying complex processes while minimizing data requirements.

Covering a 3×3 km gridded area, GDM assigns GlobeLand30 land categories to each grid and employs daily temperature and precipitation data from reference stations. It determines rain or snow in grids based on a threshold temperature. Daily ice melt and snowmelt in each grid consider debris-free and debris-covered ice, alongside glacierized and glacier-free regions.

The model computes surface runoff for each grid from precipitation, snowmelt, and ice melt. The cumulative surface runoff and baseflow from all grids contribute to the total discharge, directed to the outlet using a combined flow equation involving recession coefficients.

2.3 Recurrent neural networks (RNNs)

Recurrent neural networks (RNNs) are capable of capturing nonlinear connections in sequential data. They process input sequences element by element, utilizing a hidden state to retain information about past inputs, allowing predictions based on historical context. RNNs are suitable for regression problems, predicting continuous outputs from input sequences. Training involves employing regression loss functions like mean-squared error or mean absolute error (Heaton et al., 2018).

RNN components include an input layer, recurrent unit, hidden state, activation function, output layer, loss function, and optimization algorithm. Hyperparameters in RNNs control network behavior, impacting the capacity, convergence speed, overfitting, sequence length, dropout, layers, activation function, optimization algorithm, learning rate decay, and early stopping (Heaton et al., 2018). Types of RNNs include the simple RNN, LSTM, and GRU models. Simple RNN models possess limited memory for short-term sequences (Bengio et al., 1994). LSTM introduces memory cells and gates controlling data flow, enabling long-term memory retention (Hochreiter and Schmidhuber, 1997). GRU combines input and forget gates into an update gate, requiring fewer parameters than LSTM, but offering similar performance (Cho et al., 2014).

2.4 The GDM–RNN hybrid approach

The study introduces a GDM–RNN hybrid approach that aims to improve streamflow prediction by integrating GDM's physical constraints. This approach utilizes RNNs (simple RNN, GRU, and LSTM models) to forecast residual errors between observed and GDM-simulated inflows based on meteorological inputs. Given the complexity of quantifying these errors, the assumption is made that they exhibit identifiable patterns. The RNNs operate independently after GDM simulation to optimize error prediction without disturbing the physical constraints. Figure 2 outlines this hybrid process,

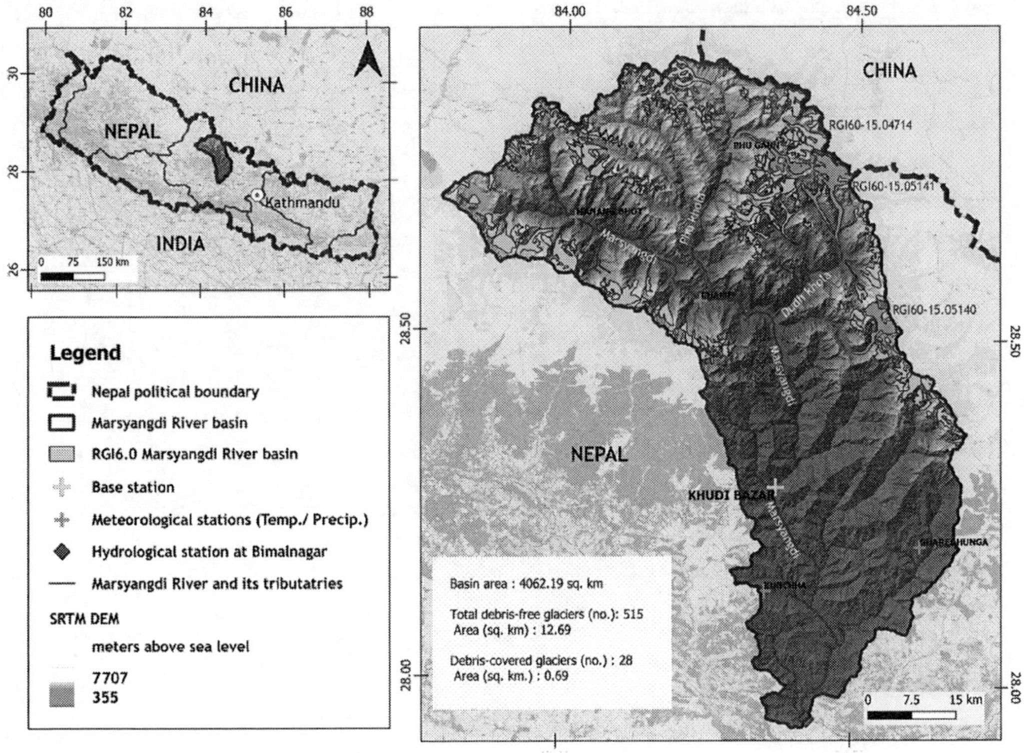

Figure 1. Map of Marsyangdi River basin, highlighting the river, tributaries, Bimalnagar hydrological station, Khudi Bazar base station, and meteorological stations. Publisher's remark: please note that the above figure contains disputed territories.

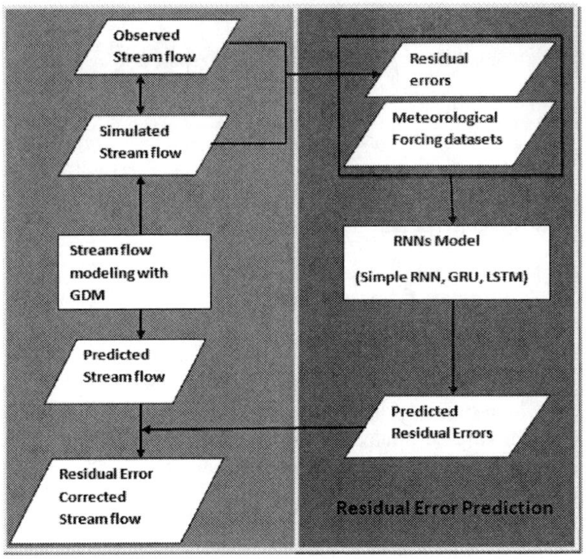

Figure 2. Schematic diagram of the GDM–RNN hybrid process.

involving the generation of simulated streamflow data via GDM, computation of residual errors, optimization, training of RNN models using these errors and meteorological data, projection of future errors, and application of these predictions to enhance GDM simulations. This approach intends to improve traditional GDM simulations by accounting for predictive discrepancies originating from various uncertainties.

2.5 Input data

Daily air temperature, rainfall, and river flow data are obtained from the Department of Hydrology and Meteorology (DHM). Temperature and rainfall data from Khudi Bazar and Chame stations are used for the GDM, while Bimalnagar outlet station's discharge data validated the study. RNNs were trained using the climatic stations' temperature, rainfall data, and discharge data from Bimalnagar. The GDM relied on elevation information from ASTER Global Digital Elevation Model Version 2 and land cover data from the GlobeLand30 dataset. Six land classes were identified and modified for uniform rainfall–runoff factors. Glacier identification was based on the International Centre for Integrated Mountain Development (ICIMOD) glacier inventory of shapefiles from 2010.

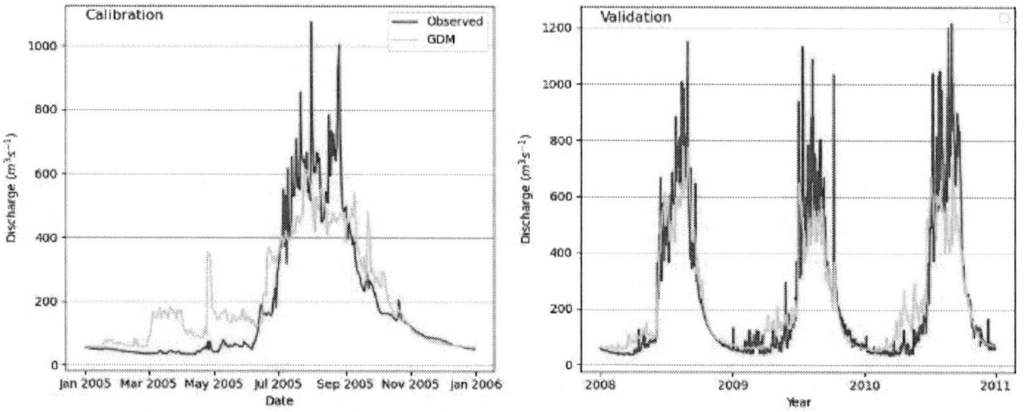

Figure 3. Observed vs. simulated discharges: the GDM model in the Marsyangdi River basin (1-year calibration and 3-year validation).

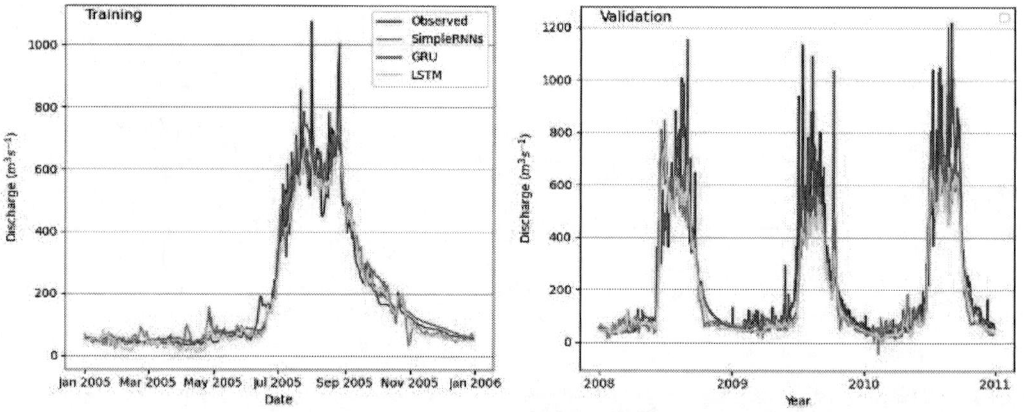

Figure 4. Observed vs. simulated discharges: different RNN models in the Marsyangdi River basin (1-year calibration and 3-year validation).

2.6 Experimental designs

The study employed seven models: GDM, GDM–LSTM, LSTM, GDM–GRU, GRU, GDM–simple RNN, and simple RNN. Each GDM model variant combined with RNNs aimed to predict residual errors. RNNs were used for river discharge simulation and performance comparisons with GDM models and GDM–RNN hybrids. Two datasets were used for calibration/training: a 1-year dataset and a 3-year dataset. GDM and RNNs were calibrated/trained using 2005 data and validated with 2008–2010 data. The GDM coupled with RNNs utilized 2005 data for calibration, predicting residual errors (2008–2010). Similarly, the experiment was repeated using data from 2005 to 2007 for calibration and training purposes. The models were then validated using 2008–2010 data. Discharge simulations by GDM (2008–2010) were corrected using predicted residual errors from RNNs, and the results were validated using observed discharge data (2008–2010).

The streamflow forecasts of various models were assessed utilizing three widely utilized measures for hydrological model evaluation: percent bias (PBIAS), Nash–Sutcliffe efficiency (NSE), and R^2 (Moriasi et al., 2015).

3 Results and discussions

In the Glacio-hydrological Degree-day Model (GDM) calibration, adjustments are made in positive degree-day factors, snow and rain coefficients, and the recession coefficient. Parameters are calibrated for 1-year and 3-year periods using observed streamflow data from the Marsyangdi River basin. The model closely replicated observed data for both periods, with slightly better performance in the 3-year calibration (Fig. 3). Although the model generally matched observed outflow, it tended to overestimate during low-flow periods due to challenges with respect to accurately depicting precipitation distribution, especially at higher altitudes (Immerzeel et al., 2015; Bocchiola et al., 2011; Barry, 2012). Evaluation

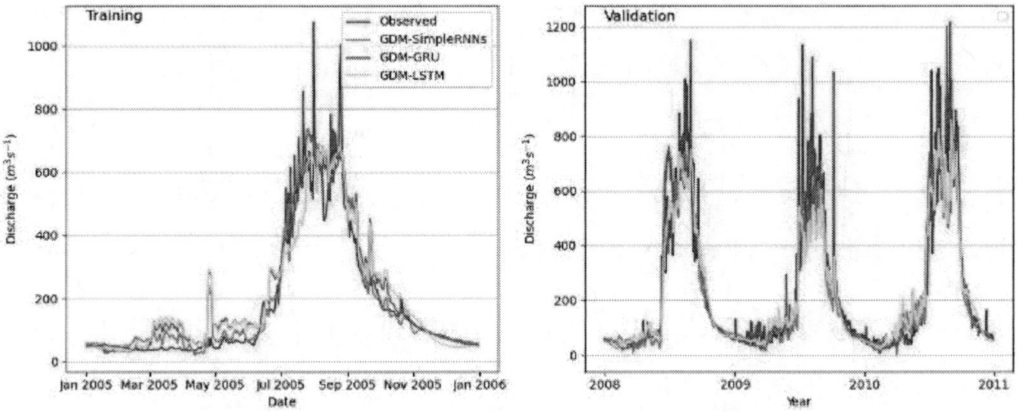

Figure 5. Observed vs. simulated discharges: different GDM–RNN models in the Marsyangdi River basin (1-year calibration and 3-year validation).

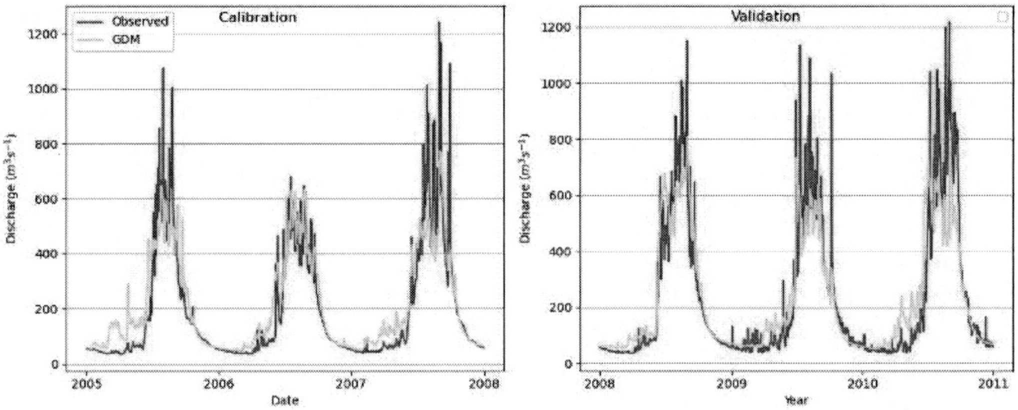

Figure 6. Observed vs. simulated discharges: the GDM model in the Marsyangdi River basin (3-year calibration and 3-year validation).

Table 1. The NSE, R^2, and PBIAS metrics during the GDM model calibration and validation periods.

Evaluation metric	One-year calibration period		Three-year calibration period	
	Calibration	Validation	Calibration	Validation
NSE	0.80	0.80	0.82	0.81
R^2	0.68	0.63	0.74	0.7
PBIAS	7.2	−4.78	10.38	−1.27

metrics (NSE, R^2, and PBIAS) indicated a satisfactory fit but showed some bias, with the model overestimating in the calibration and underestimating in the validation (Table 1).

In this study, RNN models (simple RNN, LSTM, and GRU) are trained using 1-year and 3-year datasets, employing diverse training parameters including network architec-

Table 2. The NSE, R^2, and PBIAS metrics for different RNNs used in the RNN-only approach.

Evaluation metric		One year of data used for training		Three years of data used for training	
		Training period	Validation period	Training period	Validation period
Simple RNN	NSE	0.83	0.81	0.86	0.85
	R^2	0.86	0.7	0.87	0.78
	PBIAS	10.72	−16.6	12	−8.08
LSTM	NSE	0.88	0.79	0.85	0.85
	R^2	0.88	0.65	0.88	0.82
	PBIAS	2.82	−21.42	16.99	−7.84
GRU	NSE	0.87	0.82	0.93	0.87
	R^2	0.88	0.75	0.93	0.8
	PBIAS	9.89	−12.29	9.76	−11.85

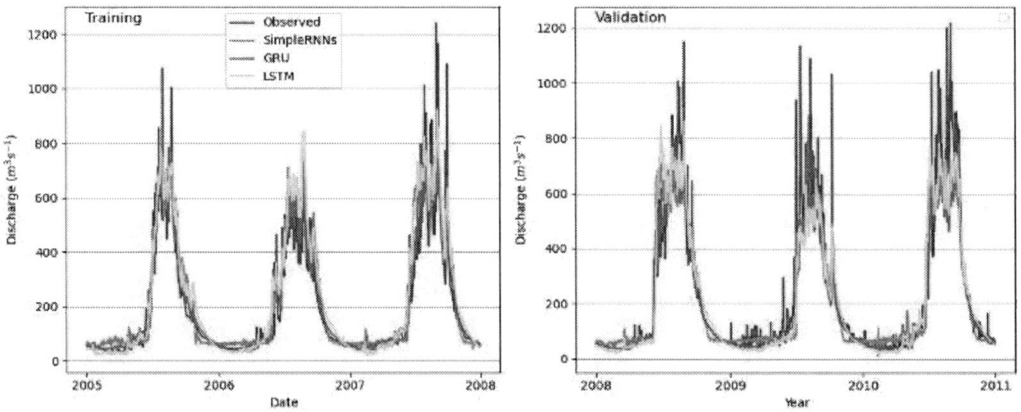

Figure 7. Observed vs. simulated discharges: different RNN models in the Marsyangdi River basin (3-year calibration and 3-year validation).

ture, activation functions, optimization algorithms, loss functions, and training schedules. Models preprocessed input data through normalization and utilized Adam optimization with a mean absolute error loss function. The number of hidden layers and neurons varied among models; all had 4 input units and 1 output unit, with GRU and LSTM using 128 neurons in a hidden layer, while simple RNN employed 32. The learning rate is constant at 0.01 but had varied schedules. In this study, the parameters of the RNN models are selected using the grid search technique.

Analysis shows superior performance for RNNs with 3 years of data compared with those with 1 year of data. LSTM and GRU exhibited better simulation of river discharge patterns than simple RNN models, especially with extensive training data, indicating a greater capacity to learn past streamflow behaviors. However, smaller datasets yielded only satisfactory results, suggesting the suitability of process-based models like GDM. Evaluation metrics demonstrated GRU's superior performance across NSE, R^2, and PBIAS for both 1-year and 3-year validation periods, while LSTM excelled slightly with respect to the R^2 value for the 3-year period. Simple RNN consistently showed the poorest performance (Table 2).

Positive PBIAS during training suggests model overestimation, potentially due to overfitting, whereas negative PBIAS during validation implies underestimation, indicating an inability to capture relevant data aspects. Despite better performance on training data, the emphasis lies on a model's ability to generalize to new data, highlighting the significance of validation results. Overall, GRU showcased superior performance in key metrics across different training periods, with LSTM showing competitive results in specific aspects.

Hybrid modeling combined GDM and three RNN variants (simple RNN, LSTM, and GRU) to predict GDM residual errors. The hybrid approach split into simulating streamflow using GDM parameters and training RNNs to predict errors (Table 3). All RNNs featured a single hidden layer with 64–254 neurons, tanh activation in the hidden layer, and linear activation in the output layer, using Glorot uniform kernel initialization. Input data normalization is applied. Key hyperparameters varied, including dropout rates (0.2–0.6), Adam optimization, batch size equivalent to the entire training set, a sequence length of 365 d, and a fixed learning rate of 0.01. Learning rate schedules varied or were absent, employing strategies like exponential or inverse time-based decay. The grid search technique is used to select the best-performing parameters.

Table 3. The NSE, R^2, and PBIAS metrics for different GDM–RNN models.

Evaluation metric		One year of data used for training		Three years of data used for training	
		Training period	Validation period	Training period	Validation period
GDM–Simple RNN	NSE	0.86	0.85	0.93	0.85
	R^2	0.88	0.82	0.93	0.78
	PBIAS	6.49	−6.21	6.16	−7.42
GDM–LSTM	NSE	0.88	0.86	0.9	0.88
	R^2	0.86	0.79	0.91	0.85
	PBIAS	6.96	−6.37	10.87	−2.22
GDM–GRU	NSE	0.87	0.85	0.85	0.85
	R^2	0.87	0.8	0.86	0.8
	PBIAS	10.4	−5.64	20.01	−0.04

The integration of GDM and RNN models (GDM–RNN) outperforms the sole use of RNNs for streamflow prediction. GDM–RNN provides more stable predictions, using residual discrepancies as objectives, while RNNs directly predict streamflow. This disparity in objectives impacts the model forecasts, potentially causing greater systematic discrepancies in RNNs due to streamflow variability.

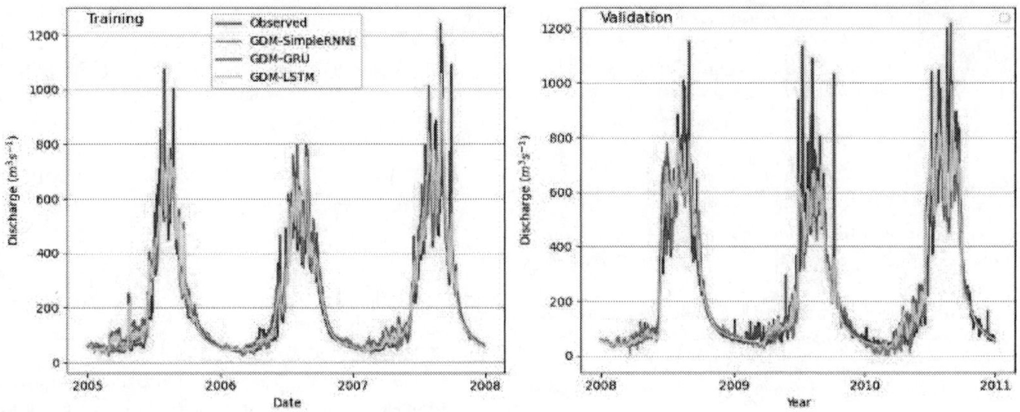

Figure 8. Observed vs. simulated discharges: different GDM–RNN models in the Marsyangdi River basin (3-year calibration)

GDM–RNN (Figs. 4–8) simulation aligns better with physical discharge patterns, addressing issues like negative discharge seen in simple RNN (Fig. 4). It particularly excels with 1 year of observed discharge data, significantly improving simulations compared with RNNs with the same training data. GDM–RNN enhances GDM's streamflow prediction by reducing uncertainties and effectively resolves high-discharge issues in the pre-monsoon season. During the monsoon, GDM–RNN (Fig. 8) closely matches observed discharge compared to GDM (Fig. 6) simulations.

4 Conclusions

Our study highlights the effectiveness of utilizing LSTM, GRU, and simple RNN machine learning techniques to improve streamflow forecasting within the GDM model in the Marsyangdi River basin, Nepal. The amalgamation of GDM and RNNs significantly enhances predictive accuracy, notably in terms of the NSE and coefficient of determination, while displaying comparable performance in PBIAS to GDM alone. The limitations of RNNs, particularly with respect to handling high-variability datasets, contribute to disparities in PBIAS measures compared with GDM. Furthermore, while GDM shows consistent performance across varying calibration dataset sizes, RNNs benefit significantly from increased training data. Notably, the GDM–RNN model displays superior adaptability with limited calibration and training data, showing notable improvements with increased dataset sizes. GRU and LSTM outperform simple RNN due to their capacity to handle long-term dependencies, both in standalone RNN usage and in hybrid combinations.

Evaluation in a single basin with limited data showcases the robustness of GDM–RNN and the potential of RNN implementation. Expanding validation with 3 years of data or across different basins would enhance its reliability. While comparing two datasets demonstrated promising results, examining additional data could further strengthen our approach.

Code availability. GDM V.2 is available upon request from Rijan Bhakta Kayastha at the Himalayan Cryosphere, Climate and Disaster Research Center (HiCCDRC), Kathmandu University.

Data availability. Hydro-meteorological data are available from the DHM, Government of Nepal (2024, http://dhm.gov.np/).

Supplement. The supplement related to this article is available online at: https://doi.org/10.5194/piahs-387-17-2024-supplement.

Author contributions. DJ developed the main idea, ran the models, and wrote the manuscript; RBK, KLS, and RK helped with data collection, the GDM simulations, and analysis.

Competing interests. At least one of the (co-)authors is a guest member of the editorial board of *Proceedings of IAHS* for the special issue "Mountain Hydrology and Cryosphere". The peer-review process was guided by an independent editor, and the authors also have no other competing interests to declare.

Special issue statement. This article is part of the special issue "Mountain Hydrology and Cryosphere". It is a result of the International Conference on Mountain Hydrology and Cryosphere, Kathmandu and Dhulikhel, Nepal, 9–10 November 2023.

References

Barry, R. G.: Recent advances in mountain climate research, Theor. Appl. Climatol., 110, 549–553, https://doi.org/10.1007/s00704-012-0695-x, 2012.

Bengio, Y., Simard, P., and Frasconi, P.: Learning Long-Term Dependencies with Gradient Descent is Difficult, IEEE Trans. Neural Netw., 5, 157–166, https://doi.org/10.1109/72.279181, 1994.

Beven, K.: Towards an alternative blueprint for a physically based digitally simulated hydrologic response modelling system, Hydrol. Process., 16, 189–206, https://doi.org/10.1002/HYP.343, 2002.

Bocchiola, D., Diolaiuti, G., Soncini, A., Mihalcea, C., D'Agata, C., Mayer, C., Lambrecht, A., Rosso, R., and Smiraglia, C.: Prediction of future hydrological regimes in poorly gauged high altitude basins: the case study of the upper Indus, Pakistan, Hydrol. Earth Syst. Sci., 15, 2059–2075, https://doi.org/10.5194/hess-15-2059-2011, 2011.

Carrera, J., Alcolea, A., Medina, A. V., Hidalgo, J. J., and Slooten, L. J.: Inverse problem in hydrogeology, Hydrogeol. J., 13, 206–222, https://doi.org/10.1007/s10040-020-02176-0, 2005.

Cho, K. and Kim, Y.: Improving streamflow prediction in the WRF-Hydro model with LSTM networks, J. Hydrol., 605, 127297, https://doi.org/10.1016/j.jhydrol.2021.127297, 2022.

Cho, K., van Merriënboer, B., Bahdanau, D., and Bengio, Y.: On the Properties of Neural Machine Translation: Encoder-Decoder Approaches, Proceedings of SSST 2014 – 8th Workshop on Syntax, Semantics and Structure in Statistical Translation, 74, 103–111, https://doi.org/10.48550/arxiv.1409.1259, 2014.

Dahlstrom, D.: Calibration and Uncertainty Analysis for Complex Environmental Models, Groundwater, 53, 1234–1245, https://doi.org/10.1111/gwat.12360, 2015.

De Filippis, G., Stevenazzi, S., Camera, C., Pedretti, D., and Masetti, M.: An agile and parsimonious approach to data management in groundwater science using open-source resources, Hydrogeol. J., 28, 1993–2008, https://doi.org/10.1007/s10040-020-02176-0, 2020.

Government of Nepal, Ministry of Energy, Water Resources and Irrigation, Department of Hydrology and Meteorology: Hydro-meteorological data, http://dhm.gov.np/, last access: 22 July 2024.

Heaton, J., Goodfellow, I., Bengio, Y., and Courville, A.: Deep learning, Genet. Program. Evolvable Mach., 19, 305–307, https://doi.org/10.1007/s10710-017-9314-z, 2018.

Hochreiter, S. and Schmidhuber, J.: Long Short-Term Memory, Neural Computat., 9, 1735–1780, https://doi.org/10.1162/NECO.1997.9.8.1735, 1997.

Immerzeel, W. W., Wanders, N., Lutz, A. F., Shea, J. M., and Bierkens, M. F. P.: Reconciling high-altitude precipitation in the upper Indus basin with glacier mass balances and runoff, Hydrol. Earth Syst. Sci., 19, 4673–4687, https://doi.org/10.5194/hess-19-4673-2015, 2015.

Ji, H., Chen, Y., Fang, G., Li, Z., Duan, W., and Zhang, Q.: Adaptability of machine learning methods and hydrological models to discharge simulations in data-sparse glaciated watersheds, J. Arid Land, 13, 549–567, https://doi.org/10.1007/s40333-021-0066-5, 2021.

Jia, X., Willard, J., Karpatne, A., Read, J. S., Zwart, J. A., Steinbach, M., and Kumar, V.: Process Guided Deep Learning for Modeling Physical Systems: An Application in Lake Temperature Modeling, International Geoscience and Remote Sensing Symposium (IGARSS), Online, 19–24 July 2020, Abstract number 3494–3496, https://doi.org/10.1109/IGARSS39084.2020.9323723, 2020.

Kayastha, R. B. and Kayastha, R.: Glacio-hydrological degree-day model (GDM) useful for the Himalayan river basins, in: Himalayan Weather and Climate and their Impact on the Environment, Springer, 379–398, https://doi.org/10.1007/978-3-030-29684-1_19, 2019.

Moriasi, D. N., Gitau, M. W., Pai, N., and Daggupati, P.: Hydrologic and Water Quality Models: Performance Measures and Evaluation Criteria, T. ASABE, 58, 1763–1785, https://doi.org/10.13031/trans.58.10715, 2015.

Nearing, G. S., Kratzert, F., Sampson, A. K., Pelissier, C. S., Klotz, D., Frame, J. M., Prieto, C., and Gupta, H. V.: What Role Does Hydrological Science Play in the Age of Machine Learning?, Water Resour. Res., 57, e2020WR028091, https://doi.org/10.1029/2020WR028091, 2020.

Refsgaard, J. C. and Knudsen, J.: Operational Validation and Intercomparison of Different Types of Hydrological Models, Water Resour. Res., 32, 2189–2202, https://doi.org/10.1029/96WR00896, 1996.

Réveillet, M., Six, D., Vincent, C., Rabatel, A., Dumont, M., Lafaysse, M., Morin, S., Vionnet, V., and Litt, M.: Relative performance of empirical and physical models in assessing the seasonal and annual glacier surface mass balance of Saint-Sorlin Glacier (French Alps), The Cryosphere, 12, 1367–1386, https://doi.org/10.5194/tc-12-1367-2018, 2018.

Solomatine, D. P.: Data-Driven Modeling and Computational Intelligence Methods in Hydrology, Encyclopedia of Hydrological Sciences, edited by: Singh, V. P., John Wiley & Sons, HSA021, https://doi.org/10.1002/0470848944.HSA021, 2006.

Proc. IAHS, 387, 25–31, 2024
https://doi.org/10.5194/piahs-387-25-2024

Open Access PIAHS

Coupling the Glacio-hydrological Degree-day Model (GDM) with PCRaster for spatial dynamic modeling of Himalayan river basins

Kundan Lal Shrestha[1], **Rijan Bhakta Kayastha**[2], **and Rakesh Kayastha**[2]

[1]Department of Chemical Science and Engineering, School of Engineering, Kathmandu University, Dhulikhel, Nepal
[2]Himalayan Cryosphere, Climate and Disaster Research Center (HiCCDRC), Department of Environmental Science and Engineering, School of Science, Kathmandu University, Dhulikhel, Nepal

Correspondence: Kundan Lal Shrestha (kundan@ku.edu.np)

Published: 18 November 2024

Abstract. We have seen a surge in glacio-hydrological modeling efforts in the past few decades. This form of modeling is also being carried out in the Himalayan river basins, but a comprehensive high-resolution simulation software that can be effective with a limited number of hydrometeorological data is recommended. In this regard, an open-source, scalable, flexible, and distributed modeling system called PyGDM has been developed by fully coupling the Glacio-hydrological Degree-day Model with PCRaster Python software. To evaluate the potential of using this model in the Himalayan river basins, we calibrated and then validated the model for Trishuli River basin using geographical data and the existing hydrometeorological data. The tests showed a promising result with respect to the effective application of the model in the entire Himalayan region. The PyGDM source code was optimized and adapted to the process models of glacier melting and hydrological processes in Himalayan basins. It increased the speed of the simulation, made the model highly scalable to accommodate new submodels, and enhanced the flexibility of the model to ingest various types of input data and parameters. Hence, the PyGDM model strives to simulate the glacio-hydrological processes of the entire Himalayan region.

1 Introduction

Modeling glacio-hydrological processes started with different modeling approaches (e.g., Buch et al., 1993; Kayastha et al., 1999) and has gained momentum, with hundreds of studies in Asia and other parts of the world focusing on glacial melting process and the related calibration and validation processes of modeling. (see the review in van Tiel et al., 2020; Y. Chen et al., 2017).

There are mainly stochastic, conceptual, and physics-based models used in semi-distributed and distributed modes (Hock and Jansson, 2006). We see some novel use of artificial neural network (ANN; Buch et al., 1993) machine learning techniques, such as long short-term memory (LSTM) and extreme gradient boosting (XGBoost) (Ji et al., 2021) in Central Asia and gated recurrent units (GRUs) integrated with the revised version of the Soil and Water Assessment Tool (SWAT+) that considers glacial hydrological processes (GSWAT+; Yang et al., 2023). Nevertheless, the ease of model development and the need for only a few types of data have given us more conceptual and semi-distributed models.

To give just a few examples of the variety of glacio-hydrological models that have been developed for various river basins around the world, the variable infiltration capacity glacier (VIC-glacier) model was used in the Eastern Pamirs (Ren et al., 2018) and China (Zhao et al., 2013); an energy-balance model (Zhang et al., 2016), the Coupled Routing and Excess Storage (CREST-glacier) model (X. Chen et al., 2017), a tracer-aided hydrological model (Nan et al., 2022), and the SWAT+ degree-day model were employed on the Tibetan Plateau (Adnan et al., 2019); the SWAT+ degree-day model (Rahman et al., 2013) and the

HBV-light model (Finger et al., 2015) were used in Switzerland; the PRMSglacier model was employed in Alaska (Van Beusekom and Viger, 2016, 2018); glacier melt models were used in the Andes (Kinouchi et al., 2013), Northwestern China (Ji and Luo, 2013), and the Upper Indus Basin (Baig et al., 2021); and the WRF-Hydro/Glacier model was employed in southern central Norway (Eidhammer et al., 2021).

The Himalayan region has also required its own set of modeling efforts (Azam et al., 2021). This has involved a mix of different modeling approaches, such as ANN models (Buch et al., 1993), the wasm overflow detector (WASMOD) model with precipitation from the Weather Research and Forecasting (WRF) model in the Beas River basin (Li et al., 2019), the VIC-glacier model in five river basins (Chandel and Ghosh, 2021), the coupled VIC–glacier melt (GM2) model (VIC-GM2) in the Tarim River (Wang et al., 2021), an energy-balance-based distributed glacier model in the Hunza River basin (Shrestha et al., 2015), a snow and glacier melt model in the Upper Indus Basin (Gupta et al., 2021), the SWAT model in the Western Himalayas (Shukla et al., 2021) and Hunza (Garee et al., 2017), a distributed hydrological model in the Western Himalayas (Wulf et al., 2016), and the snowmelt model (SNOWMOD) for the Gangotri Glacier (Arora et al., 2016).

In Nepalese Himalayan river basins, an integrated hydro-glaciological modeling system is proposed so that a common platform is available. This system has the advantage of a distributed model but only requires a few input data and parameters. This paper describes the development and testing of this model that integrates the Glacio-hydrological Degree-day Model (GDM; Kayastha et al., 2020) with PCRaster.

2 Model development

There is the need for a distributed modeling approach that is simple yet powerful and flexible to handle the vast number of spatial and temporal data generated. We chose PCRaster Python as the data processing and simulation software, as it has been successfully used in other modeling frameworks such as the Spatial Processes in Hydrology (SPHY) model (Terink et al., 2015), the PCRaster Global Water Balance 2 (PCR-GLOBWB 2) model (Sutanudjaja et al., 2017; Wiersma et al., 2022), and other similar hydrological models (Karssenberg et al., 2015; López et al., 2019).

We developed the PCRaster Python Glacio-hydrological Degree-day Model (PyGDM) by coupling the PCRaster Python framework for spatial dynamic modeling (Karssenberg et al., 2010) with the Glacio-hydrological Degree-day Model (originally developed in Fortran and then in C#) (Kayastha et al., 2005, 2020).

Our PyGDM model is a spatially distributed and parsimonious glacio-hydrological model especially designed for application in Himalayan river basins. The simplicity of the GDM is coupled with the power and ease of the dynamic modeling of PCRaster by embedding the GDM code into the initial and dynamic sections of the PCRaster script.

2.1 Modeling framework

The modeling method is shown in Fig. 1. Kayastha et al. (2003) proposed positive degree-day factors for the Himalayan river basins, and this method has become a simple and convenient approach to simulate glacierized basin discharge in the Himalayan basins (e.g., Pradhananga et al., 2014; Silwal et al., 2016; Kayastha et al., 2020).

Precipitation is calculated as rain or snow according the threshold temperature (Eq. 1):

$$\text{Precipitation} = \begin{cases} \text{rain} & T \geq T_\mathrm{T} \\ \text{snow} & T < T_\mathrm{T}, \end{cases} \quad (1)$$

where T is temperature and T_T is the critical/threshold temperature.

The snowmelt and ice melt value, M (mm d^{-1}), is calculated using degree-day factors for snow, bare ice, and debris-covered ice:

$$M = \begin{cases} K_\mathrm{d} \text{ or } K_\mathrm{s} \text{ or } K_\mathrm{b} \times T & T > 0 \\ 0 & T \leq 0, \end{cases} \quad (2)$$

where T is air temperature (°C), K_s is the degree-day factor for snow (mm °C d^{-1}), K_b is the degree-day factor for bare ice (mm °C d^{-1}), and K_d is the degree-day factor for debris-covered ice (mm °C d^{-1}).

The Hargreaves method is a very simple empirical method for estimating potential evapotranspiration (PET) based on temperature data. The advantage of using this method is that we need only temperature data as input; this enables us to generalize the method for the entire Nepalese domain. The Hargreaves method estimates PET based on Eq. (3):

$$\mathrm{PET} = 0.0023\,(T_\mathrm{mean} + 17.8)\,R_a\,0.0408\,\sqrt{T_\mathrm{max} - T_\mathrm{min}}, \quad (3)$$

where PET is potential evapotranspiration (mm d^{-1}); T_mean is average daily temperature (°C); T_max is maximum daily temperature (°C); T_min is minimum daily temperature (°C); and R_a is extraterrestrial radiation (MJ m^2 d^{-1}), which can be calculated based on location, date, and time using established formulas.

The Hargreaves method is also easy to use due to its dependence on only temperature and radiation, although we may consider other variables such as humidity and wind in future versions of the model. A simple interception model was used with monthly threshold values provided in the model.

The two-reservoir-based modeling approach of SWAT (Luo et al., 2012) is adopted to simulate the hydrological response of the baseflow and rainfall runoff contribution to river discharge. Then, the discharge due to rainfall and snowfall are calculated with the use of runoff coefficients, and the

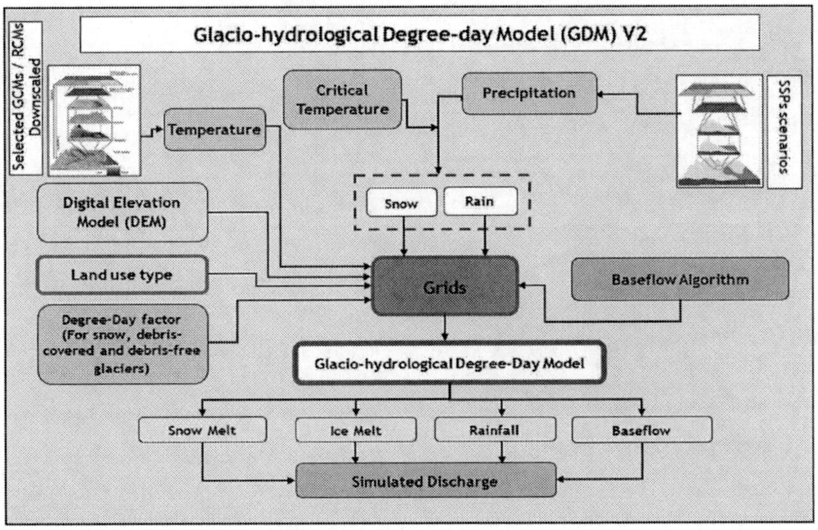

Figure 1. PyGDM modeling framework. The abbreviations used in the figure are as follows: GCM – global circulation model; RCM – regional climate model; and SSP – Shared Socioeconomic Pathway.

daily accumulated amount of water flowing through the river basin is simulated using the partial contribution from the past-day discharge (Eqs. 4, 5). The details of the discharge simulation are explained in more precisely in Kayastha et al. (2020).

We decided to utilize the static and dynamic modeling capabilities of PCRaster software to simulate the hydrological response of the Himalayan river basins. PCRaster is a highly capable environmental modeling software with a Python version that is open-source and free. The deterministic modeling required for the distributed hydrological modeling starts with the static modeling framework, as summarized by Eq. (6). In Eq. (6), Z is the model state variable, I represents the inputs, P represents the parameters, and f defines the model structure. In the dynamic modeling section of the PCRaster script, a time component is added in the static model. In the dynamic modeling framework, there are several sections such as the "initial" section, the "dynamic" section, the dynamic database, the timer, and so on. The initial section is used to initialize static values and parameters used in the hydrological model, to store lookup tables of various parameters, and to define the major geographical databases, such as digital elevation model, land use and land cover, and basin spatial maps. The dynamic section is used to apply the temporal behavior of the hydrological model by running the operations on maps for every time step in sequence. PCRaster also has a multicore module to facilitate the parallel running mode. We used the multicore capability in the workstation to decrease the simulation time.

$$Q_G = Q_r \cdot C_r + Q_s \cdot C_s \tag{4}$$

$$Q_d = Q_R(1-k) + Q_{R(d-1)}k + Q_B \tag{5}$$

$$Z = f(Z, I, P) \tag{6}$$

The only spatially and temporally varying input data required for our model are the precipitation and temperature. Our model is able to interpolate the station data as well as use the gridded NetCDF data files that may be obtained from global and regional models and reanalysis data.

2.2 Input data

In the present version of our model, land use and land cover are classified into eight categories: forest, grassland, agriculture, barren land, waterbodies, built-up area, debris-covered glacier, and clean/bare glacier.

To calibrate and validate PyGDM, we applied the model to the Trishuli River basin, which covers part of China and Nepal. We used the gridded land use and land cover data (Fig. 2), which were obtained by modifying and reclassifying the Esri LULC (land use and land cover) derived from Sentinel-2 imagery at a 10 m resolution (Karra et al., 2021) (https://livingatlas.arcgis.com/landcoverexplorer, last access: 10 June 2024). To reclassify land classes for glacier, the Randolph Glacier Inventory Version 6.0 (RGI v6.0) glacier outlines were used. A 30 m resolution digital elevation model (DEM) was obtained from the Shuttle Radar Topography Mission (SRTM) 1 arcsec Global 30 m DEM from the United States Geological Survey (https://earthexplorer.usgs.gov, last access: 10 June 2024).

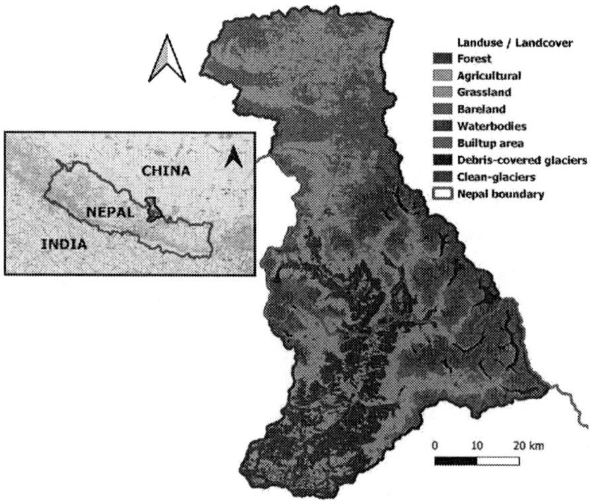

Figure 2. Land use and land cover (LULC) of Trishuli River basin. We used the LULC from Esri and modified it as per the requirements of our model. The Esri LULC was produced by Impact Observatory, Microsoft, and Esri (Karra et al., 2021). Publisher's remark: please note that the above figure contains disputed territories.

Table 1. Parameters and coefficients for modeling.

Parameters	Values
$\delta_{gw,sh}$ (d)	10
$\delta_{gw,dp}$ (d)	65
$\alpha_{gw,sh}$	0.3
$\alpha_{gw,dp}$	0.5
β_{dp}	0.8
K_x	0.7
T_c (°C)	2
k_d	3

2.3 Parameters

Table 1 shows the main parameters used in the discharge simulation, baseflow calculation of shallow and deep aquifers, and critical temperature. All of the parameters have been identified from our past research (Kayastha et al., 2020).

Table 2 shows the degree-day factors, runoff coefficients, and interception thresholds for each of the land use categories used in the model. These factors and coefficients are taken from our research as well as past publications (Kayastha et al., 2005, 2020).

The interception is dependent on the land use and is modeled as a threshold value. The interception thresholds vary per land use type and have been adopted from the guidelines provided by Gerrits (2010).

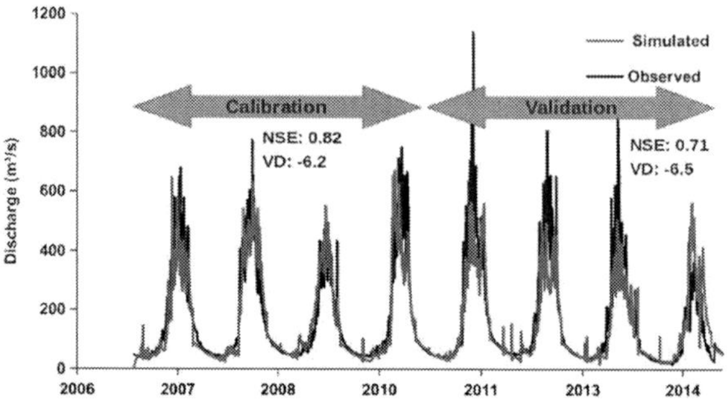

Figure 3. Calibration and validation of discharge for the Trishuli River basin.

3 Results and discussion

The PyGDM model ran successfully in multicore mode with highly acceptable results. The simulation time was below 10 min for a 1-year simulation at a 30 m grid resolution. The simulation time period may be altered by the actual computational power, RAM, and the CPU core characteristics.

A total of 4 years of simulation was used for the calibration and validation of the model, respectively. The calibration and validation results of the model are shown in Fig. 3. The Nash–Sutcliffe efficiency and the volume difference were the statistical metrics used in the evaluation. Both of the metrics are within the acceptable limits of the performance of the models. PyGDM is also capable of evaluating the component contribution of hydrological processes, such as baseflow, rain, snowmelt, and ice melt, on the total discharge of the basin. The contribution of process components to the discharge of Trishuli River basin is shown in Fig. 4. Although the spatial variation is not presented in this paper, the PyGDM modeling system is able to plot the spatially distributed maps of the output variables as well as animations depicting the temporal development of the hydrological response of the basin. The calibration and validation results are within the acceptable regime. The speed of simulation in multicore mode also promises very effective application of this model in multiyear and decadal simulations of the hydrological response and climate change scenarios in the Himalayan river basins. PyGDM is also capable of ingesting a diverse set of input data, and the use of the Python language gives an easy platform to upscale the model with respect to the use of new subprocesses.

4 Conclusions

A glacio-hydrological model built on positive degree-day factors for glacier melt modeling was successfully integrated with PCRaster Python software to enable a distributed approach and enhance the scalability of the conceptual model for application in the entire Himalayan region. The dynamic

Table 2. Degree-day factors and runoff coefficients.

Month	Degree-day factor		Runoff coefficient for eight land uses							
	K_b	K_s	Rc_L1	Rc_L2	Rc_L3	Rc_L4	Rc_L5	Rc_L6	Rc_L7	Rc_L8
Jan	10.5	9.5	0.3	0.5	0.5	0.3	0.95	0.95	1	1
Feb	10.5	9.5	0.3	0.5	0.5	0.3	0.95	0.95	1	1
Mar	10.5	9.5	0.3	0.5	0.5	0.3	0.95	0.95	1	1
Apr	10.5	9.5	0.3	0.5	0.5	0.3	0.95	0.95	1	1
May	10.5	9.5	0.3	0.5	0.5	0.3	0.95	0.95	1	1
Jun	8.0	8	0.7	0.8	0.7	0.7	0.95	0.95	1	1
Jul	8.0	8	0.7	0.8	0.7	0.7	0.95	0.95	1	1
Aug	8.0	8	0.7	0.8	0.7	0.7	0.95	0.95	1	1
Sep	8.0	8	0.7	0.8	0.7	0.7	0.95	0.95	1	1
Oct	10.5	9.5	0.3	0.5	0.5	0.3	0.95	0.95	1	1
Nov	10.5	9.5	0.3	0.5	0.5	0.3	0.95	0.95	1	1
Dec	10.5	9.5	0.3	0.5	0.5	0.3	0.95	0.95	1	1

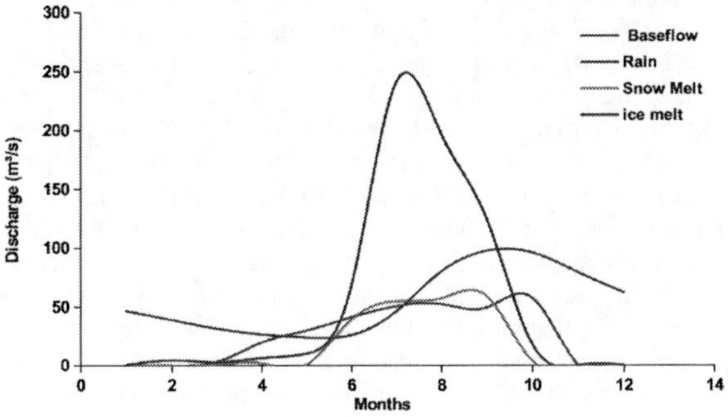

Figure 4. Components of discharge for the Trishuli River basin.

modeling capability of PCRaster has been coupled with the simple yet highly effective glacio-hydrological model so that the spatiotemporal resolution of the simulation could be enhanced. To test this newly developed PyGDM model, calibration and validation phases of model development were successfully completed for the Trishuli River basin. PyGDM is ready for the comprehensive modeling and simulation of glacio-hydrological processes in the Himalayan basins. The flexibility, scalability, and the simplicity of using available environmental data in the Himalayan region are the main advantages of PyGDM.

Code availability. The code for the PyGDM model is available upon request.

Data availability. The data supporting the findings of this study are available from the corresponding author upon request.

Author contributions. KLS: model coding and development, original draft, methodology, and editing. RBK: GDM concept, supervision, and review and editing. RK: simulation results and plotting figures.

Competing interests. At least one of the (co-)authors is a guest member of the editorial board of *Proceedings of IAHS* for the special issue "Mountain Hydrology and Cryosphere". The peer-review process was guided by an independent editor, and the authors also have no other competing interests to declare.

Disclaimer. Publisher's note: Copernicus Publications remains neutral with regard to jurisdictional claims made in the text, published maps, institutional affiliations, or any other geographical representation in this paper. While Copernicus Publications makes every effort to include appropriate place names, the final responsibility lies with the authors.

Special issue statement. This article is part of the special issue "Mountain Hydrology and Cryosphere". It is a result of the International Conference on Mountain Hydrology and Cryosphere, Kathmandu and Dhulikhel, Nepal, 9–10 November 2023.

Financial support. This research has been supported by the University Grants Commission Nepal (grant no. UGC COVID-19 Response, Technical Innovation, Scientific Investigation and Research Grant 2077/78).

References

Adnan, M., Kang, S.-c., Zhang, G.-s., Anjum, M. N., Zaman, M., and Zhang, Y.-q.: Evaluation of SWAT Model performance on glaciated and non-glaciated subbasins of Nam Co Lake, Southern Tibetan Plateau, China, J. Mt. Sci., 16, 1075–1097, https://doi.org/10.1007/s11629-018-5070-7, 2019.

Arora, M., Kumar, R., Kumar, N., and Malhotra, J.: Hydrological Modeling and Streamflow Characterization of Gangotri Glacier, in: Geostatistical and Geospatial Approaches for the Characterization of Natural Resources in the Environment, edited by: Raju, N., Springer, Cham., 909–915, https://doi.org/10.1007/978-3-319-18663-4_141, 2016.

Azam, M. F., Kargel, J. S., Shea, J. M., Nepal, S., Haritashya, U. K., Srivastava, S., Maussion, F., Qazi, N., Chevallier, P., Dimri, A. P., Kulkarni, A. V., Cogley, J. G., and Bahuguna, I.: Glaciohydrology of the Himalaya-Karakoram, Science, 373, eabf3668, https://doi.org/10.1126/science.abf3668, 2021.

Baig, S., Sayama, T., and Takara, K.: Hydrological Modeling of the Astore River Basin, Pakistan, by Integrating Snow and Glacier Melt Processes and Climate Scenarios, Journal of Disaster Research, 16, 1197–1206, https://doi.org/10.20965/jdr.2021.p1197, 2021.

Buch, A., Mazumdar, H., and Pandey, P.: Application of artificial neural networks in hydrological modeling: a case study of runoff simulation of a Himalayan glacier basin, Proceedings of 1993 International Conference on Neural Networks (IJCNN-93-Nagoya, Japan), 25–29 October 1993, Nagoya, Japan, https://doi.org/10.1109/ijcnn.1993.714073, 1993.

Chandel, V. S. and Ghosh, S.: Components of Himalayan River Flows in a Changing Climate, Water Resour. Res., 57, e2020WR027589, https://doi.org/10.1029/2020wr027589, 2021.

Chen, X., Long, D., Hong, Y., Zeng, C., and Yan, D.: Improved modeling of snow and glacier melting by a progressive two-stage calibration strategy with GRACE and multisource data: How snow and glacier meltwater contributes to the runoff of the Upper Brahmaputra River basin?, Water Resour. Res., 53, 2431–2466, https://doi.org/10.1002/2016wr019656, 2017.

Chen, Y., Li, W., Fang, G., and Li, Z.: Review article: Hydrological modeling in glacierized catchments of central Asia – status and challenges, Hydrol. Earth Syst. Sci., 21, 669–684, https://doi.org/10.5194/hess-21-669-2017, 2017.

Eidhammer, T., Booth, A., Decker, S., Li, L., Barlage, M., Gochis, D., Rasmussen, R., Melvold, K., Nesje, A., and Sobolowski, S.: Mass balance and hydrological modeling of the Hardangerjøkulen ice cap in south-central Norway, Hydrol. Earth Syst. Sci., 25, 4275–4297, https://doi.org/10.5194/hess-25-4275-2021, 2021.

Finger, D., Vis, M., Huss, M., and Seibert, J.: The value of multiple data set calibration versus model complexity for improving the performance of hydrological models in mountain catchments, Water Resour. Res., 51, 1939–1958, https://doi.org/10.1002/2014wr015712, 2015.

Garee, K., Chen, X., Bao, A., Wang, Y., and Meng, F.: Hydrological Modeling of the Upper Indus Basin: A Case Study from a High-Altitude Glacierized Catchment Hunza, Water, 9, 17, https://doi.org/10.3390/w9010017, 2017.

Gerrits, A.: The role of interception in the hydrological cycle, PhD thesis, Civil Engineering and Geosciences, VSSD, TU Delft, ISBN 978-90-6562-248-8, 2010.

Gupta, C., Kulkarni, A. V., and Taloor, A. K.: Streamflow modeling and contribution of snow and glacier melt runoff in glacierized Upper Indus Basin, Environ. Monit. Assess., 193, 761, https://doi.org/10.1007/s10661-021-09537-6, 2021.

Hock, R. and Jansson, P.: Modeling Glacier Hydrology, chap. 170, John Wiley & Sons, Ltd, ISBN 9780470848944, https://doi.org/10.1002/0470848944.hsa176, 2006.

Ji, H., Chen, Y., Fang, G., Li, Z., Duan, W., and Zhang, Q.: Adaptability of machine learning methods and hydrological models to discharge simulations in data-sparse glaciated watersheds, J. Arid Land, 13, 549–567, https://doi.org/10.1007/s40333-021-0066-5, 2021.

Ji, X. and Luo, Y.: The influence of precipitation and temperature input schemes on hydrological simulations of a snow and glacier melt dominated basin in Northwest China, Hydrol. Earth Syst. Sci. Discuss., 10, 807–853, https://doi.org/10.5194/hessd-10-807-2013, 2013.

Karra, K., Kontgis, C., Statman-Weil, Z., Mazzariello, J. C., Mathis, M., and Brumby, S. P.: Global land use/land cover with Sentinel 2 and deep learning, in: 2021 IEEE International Geoscience and Remote Sensing Symposium IGARSS, 2021–2021, IEEE, 11–16 July 2021, Brussels, Belgium, https://doi.org/10.1109/igarss47720.2021.9553499, 2021.

Karssenberg, D., Schmitz, O., Salamon, P., de Jong, K., and Bierkens, M. F.: A software framework for construction of process-based stochastic spatio-temporal models and data assimilation, Environ. Modell. Softw., 25, 489–502, https://doi.org/10.1016/j.envsoft.2009.10.004, 2010.

Karssenberg, D., Lana-Renault, N., and Schmitz, O.: PyCatch: catchment modelling in the PCRaster framework, Geophys. Res. Abstr., EGU2015-9327, EGU General Assembly 2015, Vienna, Austria, https://ui.adsabs.harvard.edu/abs/2015EGUGA..17.9327K (last access: 10 June 2024), 2015.

Kayastha, R. B., Ohata, T., and Ageta, Y.: Application of a mass-balance model to a Himalayan glacier, J. Glaciol., 45, 559–567, https://doi.org/10.1017/s002214300000143x, 1999.

Kayastha, R. B., Ageta, Y., Nakawo, M., Fujita, K., Sakai, A., and Matsuda, Y.: Positive degree-day factors for ice ablation on four glaciers in the Nepalese Himalayas and Qinghai-Tibetan Plateau, Bulletin of Glaciological Research, 20, 7–14, 2003.

Kayastha, R. B., Ageta, Y., and Fujita, K.: Use of Positive Degree-Day Methods for Calculating Snow and Ice Melting and Discharge in Glacierized Basins in the Langtang Valley, Central Nepal, chap. 2, 5–14, John Wiley & Sons, Ltd, ISBN 9780470858240, https://doi.org/10.1002/0470858249.ch2, 2005.

Kayastha, R. B., Steiner, N., Kayastha, R., Mishra, S. K., and McDonald, K.: Comparative Study of Hydrology and Icemelt in Three Nepal River Basins Using the Glacio-Hydrological Degree-Day Model (GDM) and Observations From the Advanced Scatterometer (ASCAT), Front. Earth Sci., 7, 354, https://doi.org/10.3389/feart.2019.00354, 2020.

Kinouchi, T., Liu, T., Mendoza, J., and Asaoka, Y.: Modeling glacier melt and runoff in a high-altitude headwater catchment in the Cordillera Real, Andes, Hydrol. Earth Syst. Sci.

Discuss., 10, 13093–13144, https://doi.org/10.5194/hessd-10-13093-2013, 2013.

Li, L., Shen, M., Hou, Y., Xu, C.-Y., Lutz, A. F., Chen, J., Jain, S. K., Li, J., and Chen, H.: Twenty-first-century glacio-hydrological changes in the Himalayan headwater Beas River basin, Hydrol. Earth Syst. Sci., 23, 1483–1503, https://doi.org/10.5194/hess-23-1483-2019, 2019.

Luo, Y., Arnold, J., Allen, P., and Chen, X.: Baseflow simulation using SWAT model in an inland river basin in Tianshan Mountains, Northwest China, Hydrol. Earth Syst. Sci., 16, 1259–1267, https://doi.org/10.5194/hess-16-1259-2012, 2012.

López, N. A., Santos, G., and Perez, G. A. C.: Development of a distributed hydrological model for the Coello river basin in Colombia, in: 38th IAHR World Congress – "Water: Connecting the World", 38IAHRWC, The International Association for Hydro-Environment Engineering and Research (IAHR), 1–6 September 2019, Panama City, Panama, ISSN 2521-7119, https://doi.org/10.3850/38wc092019-1825, 2019.

Nan, Y., He, Z., Tian, F., Wei, Z., and Tian, L.: Assessing the influence of water sampling strategy on the performance of tracer-aided hydrological modeling in a mountainous basin on the Tibetan Plateau, Hydrol. Earth Syst. Sci., 26, 4147–4167, https://doi.org/10.5194/hess-26-4147-2022, 2022.

Pradhananga, N. S., Kayastha, R. B., Bhattarai, B. C., Adhikari, T. R., Pradhan, S. C., Devkota, L. P., Shrestha, A. B., and Mool, P. K.: Estimation of discharge from Langtang River basin, Rasuwa, Nepal, using a glacio-hydrological model, Ann. Glaciol., 55, 223–230, https://doi.org/10.3189/2014aog66a123, 2014.

Rahman, K., Maringanti, C., Beniston, M., Widmer, F., Abbaspour, K., and Lehmann, A.: Streamflow Modeling in a Highly Managed Mountainous Glacier Watershed Using SWAT: The Upper Rhone River Watershed Case in Switzerland, Water Resour. Manag., 27, 323–339, https://doi.org/10.1007/s11269-012-0188-9, 2013.

Ren, Z., Su, F., Xu, B., Xie, Y., and Kan, B.: A Coupled Glacier-Hydrology Model and Its Application in Eastern Pamir, J. Geophys. Res.-Atmos., 123, 13692–13713, https://doi.org/10.1029/2018jd028572, 2018.

Shrestha, M., Koike, T., Hirabayashi, Y., Xue, Y., Wang, L., Rasul, G., and Ahmad, B.: Integrated simulation of snow and glacier melt in water and energy balance-based, distributed hydrological modeling framework at hunza river basin of Pakistan Karakoram region, J. Geophys. Res., 120, 4889–4919, https://doi.org/10.1002/2014jd022666, 2015.

Shukla, S., Jain, S. K., and Kansal, M. L.: Hydrological modelling of a snow/glacier-fed western Himalayan basin to simulate the current and future streamflows under changing climate scenarios, Sci. Total Environ., 795, 148871, https://doi.org/10.1016/j.scitotenv.2021.148871, 2021.

Silwal, G., Kayastha, R. B., and Mool, P. K.: Application of Temperature Index Model for Estimating Daily Discharge of Sangda River Basin, Mustang, Nepal, J. Comput. Chem., 2, 15–26, https://doi.org/10.3233/jcc-160002, 2016.

Sutanudjaja, E. H., van Beek, R., Wanders, N., Wada, Y., Bosmans, J. H. C., Drost, N., van der Ent, R. J., de Graaf, I. E. M., Hoch, J. M., de Jong, K., Karssenberg, D., López López, P., Peßenteiner, S., Schmitz, O., Straatsma, M. W., Vannametee, E., Wisser, D., and Bierkens, M. F. P.: PCR-GLOBWB 2: a 5 arcmin global hydrological and water resources model, Geosci. Model Dev., 11, 2429–2453, https://doi.org/10.5194/gmd-11-2429-2018, 2018.

Terink, W., Lutz, A. F., Simons, G. W. H., Immerzeel, W. W., and Droogers, P.: SPHY v2.0: Spatial Processes in HYdrology, Geosci. Model Dev., 8, 2009–2034, https://doi.org/10.5194/gmd-8-2009-2015, 2015.

Van Beusekom, A. E. and Viger, R. J.: A Glacier Runoff Extension to the Precipitation Runoff Modeling System, J. Geophys. Res., 121, 2001–2021, https://doi.org/10.1002/2015jf003789, 2016.

Van Beusekom, A. E. and Viger, R. J.: A Physically Based Daily Simulation of the Glacier-Dominated Hydrology of the Copper River Basin, Alaska, Water Resour. Res., 54, 4983–5000, https://doi.org/10.1029/2018wr022625, 2018.

van Tiel, M., Stahl, K., Freudiger, D., and Seibert, J.: Glacio-hydrological model calibration and evaluation, WIREs Water, 7, e1483, https://doi.org/10.1002/wat2.1483, 2020.

Wang, N., Liu, W., Wang, H., Sun, F., Duan, W., Li, Z., Li, Z., and Chen, Y.: Improving streamflow and flood simulations in three headwater catchments of the Tarim River based on a coupled glacier-hydrological model, J. Hydrol., 603, 127048, https://doi.org/10.1016/j.jhydrol.2021.127048, 2021.

Wiersma, P., Aerts, J., Zekollari, H., Hrachowitz, M., Drost, N., Huss, M., Sutanudjaja, E. H., and Hut, R.: Coupling a global glacier model to a global hydrological model prevents underestimation of glacier runoff, Hydrol. Earth Syst. Sci., 26, 5971–5986, https://doi.org/10.5194/hess-26-5971-2022, 2022.

Wulf, H., Bookhagen, B., and Scherler, D.: Differentiating between rain, snow, and glacier contributions to river discharge in the western Himalaya using remote-sensing data and distributed hydrological modeling, Adv. Water Resour., 88, 152–169, https://doi.org/10.1016/j.advwatres.2015.12.004, 2016.

Yang, C., Xu, M., Kang, S., Fu, C., and Hu, D.: Improvement of streamflow simulation by combining physically hydrological model with deep learning methods in data-scarce glacial river basin, J. Hydrol., 625, 129990, https://doi.org/10.1016/j.jhydrol.2023.129990, 2023.

Zhang, G., Kang, S., Cuo, L., and Qu, B.: Modeling hydrological process in a glacier basin on the central Tibetan Plateau with a distributed hydrology soil vegetation model, J. Geophys. Res., 121, 9521–9539, https://doi.org/10.1002/2016jd025434, 2016.

Zhao, Q., Ye, B., Ding, Y., Zhang, S., Yi, S., Wang, J., Shangguan, D., Zhao, C., and Han, H.: Coupling a glacier melt model to the Variable Infiltration Capacity (VIC) model for hydrological modeling in north-western China, Environ. Earth Sci., 68, 87–101, https://doi.org/10.1007/s12665-012-1718-8, 2013.

Proc. IAHS, 387, 33–39, 2024
https://doi.org/10.5194/piahs-387-33-2024

Open Access PIAHS

Potential of tree-ring chronologies for multi-centennial streamflow reconstructions: an insight from Nepal

Narayan P. Gaire[1], **Yub R. Dhakal**[1], **Santosh K. Shah**[2], **and Ze-Xin Fan**[3]

[1]Department of Environmental Science, Patan Multiple Campus, Tribhuvan University, Patan Dhoka, Lalitpur, Nepal
[2]Birbal Sahni Institute of Palaeosciences, Lucknow 226 007, India
[3]CAS Key Laboratory of Tropical Forest Ecology, Xishuangbanna Tropical Botanical Garden, Chinese Academy of Sciences, Menglun, Mengla, Yunnan 666303, China

Correspondence: Narayan P. Gaire (npgaire2007@gmail.com)

Published: 18 November 2024

Abstract. Rivers in the Himalayan and adjacent mountain regions are the lifelines of over 1 billion people and are the backbone of civilizations therein. The short gauge records of Nepal do not provide a sufficient time window to understand the natural variations in river discharge from a long-term climate perspective. By developing a network of over 100 tree-ring chronologies across Nepal, we checked their hydrological sensitivity for long-term streamflow reconstruction. This shows huge potential for long-term annual or seasonal streamflow reconstructions in different river basins of Nepal. A robust reconstruction model was developed between tree growth and streamflow, capturing 56 % of the variance in the actual data, and was used to reconstruct the March–July monthly average streamflow of Sinja Khola (river) at Diware from AD 1700 to 2013. The reconstruction revealed several dry and pluvial periods with the recent decline in the streamflow in the Sinja River. We found short- (2 to 8.7 years) to medium-term (35.2 years) periodicities in the reconstruction that are likely to be associated with climatic oscillations, such as the El Niño–Southern Oscillation (ENSO), the Atlantic Multidecadal Oscillation (AMO), and the North Atlantic Oscillation (NAO), along with the influence of local circulation patterns. Since Sinja Valley is related to the origin of the Nepali language and civilization, the information on long-term streamflow will be beneficial for water resource management in the context of rapid climate change and for preparedness for water-induced disasters in the region.

1 Introduction

Climate change is an essential issue for the Hindu Kush Himalaya (HKH) region, which is witnessing a rapid increase in temperature, one that is more significant than the global average, along with increasing extremes in temperature and precipitation (DHM, 2017; Talchabhadel et al., 2018; ICIMOD, 2023). The ongoing climate change has left several bio-physical imprints in the HKH region (Chaudhary et al., 2023; Gaire et al., 2023; ICIMOD, 2023). The Himalayas and the adjacent mountains, the water towers of Asia, provide different ecosystem services and benefit more than 2 billion people in the river basins of the region (Immerzeel et al., 2010; ICIMOD, 2023); they are being severely affected by rapid climate change in the area. Research pointed out a marked reduction in water availability due to the shrinking of glaciers (Bolch et al., 2012; ICIMOD, 2023). However, modelling studies indicated an increasing river flow in the HKH rivers until the mid-21st century (Mathison et al., 2015).

Nepal harbours over 6000 rivers and rivulets, with the Karnali, Gandaki, and Koshi being three major river systems. The short river gauge station datasets in Nepal create a gap in understanding the long-term variations in river discharge. The long-term quantitative hydro-climatic variations are imperative for water resource management for different sectors, including agriculture, water and energy resources, biodiversity conservation, accurate design of infrastructures, and formulation of appropriate adaptation measures against climate change impacts and other disasters.

Nepal has several annual tree-ring-producing moisture-sensitive tree taxa, which are sources of high-resolution climate proxy information for long-term hydro-climatic reconstructions (Gaire et al., 2017, 2019; Liang et al., 2019; Panthi et al., 2017). Reconstructions of the magnitude, intensity, and periodicity of streamflow variability at different temporal scales provide a robust basis for planning water resources for recently implemented hydropower and irrigation projects in Nepal. Therefore, this study aims to provide the hydrological reconstruction potential of existing tree-ring chronology networks from Nepal and to reconstruct multi-centennial long streamflows in Nepal's rivers.

2 Materials and methods

2.1 Hydrological data

Available hydrological data closer to the tree-ring sampling sites were collected from the Department of Hydrology and Meteorology, Government of Nepal.

2.2 Collection and analysis of tree-ring data

Tree-ring data were collected from both primary and secondary sources. All publicly available existing tree-ring chronologies were collected from public repositories: the International Tree-Ring Data Bank (ITRDB) and the National Oceanic and Atmospheric Administrations (https://www.ncei.noaa.gov/products/paleoclimatology/tree-ring, last access: 11 March 2019). In addition, new tree-ring chronologies were developed from our sample collection. Tree core samples were analysed in the lab following standard procedure (Fritts, 1976). Each ring of all the core samples was counted, dated, and measured and cross-dated using the LinTab™ measuring system attached to a computer with the TSAP-Win package (Rinn, 2013). The accuracy of cross-dating and measurement was further checked using the COFECHA programme. Standardization in ARSTAN and dplR (Bunn, 2008) was done for raw tree-ring width series to remove the trend in the raw chronology related to the tree age and geometry and to maximize the common signal likely to be associated with climate (Fritts, 1976). The various chronology statistics were calculated to assess the quality and dendroclimatic potential of the developed individual site or composite chronologies (Fritts, 1976; Wigley et al., 1984). Composite chronologies were created by combining the individual site chronologies showing similar growth trends and climatic responses or by combining chronologies of the same river catchments.

2.3 Hydro-climate sensitivity and response analysis and long-streamflow reconstruction

To establish growth–climate relationships and to identify growth-limiting climatic factors, simple and bootstrapped correlation analyses were carried out between hydro-climate (temperature, precipitation, and streamflow) and tree-ring chronology (Zang and Biondi, 2015). Later, a significant response was used to develop a regression model for the reconstruction of the past hydro-climate of the target variable and target season. For this study, we presented the spring–early-summer season (March–July) streamflow reconstruction of Sinja Khola (a tributary of Karnali River) at Diware station in Jumla as an example of the application of tree rings for long-term streamflow studies. The reconstruction was verified using leave-one-out cross-validation (Michaelsen, 1987) using various calibration–verification statistics (Fritts, 1976; Cook and Kairiukstis, 1990). The reconstruction was truncated at the point at which the EPS value became less than the arbitrary but commonly used threshold value (Wigley et al., 1984). Power spectral analysis using the multi-tapered method and the Morlet global wavelet analysis was performed to detect and analyse periodicities of the reconstructed streamflow series and to establish possible teleconnection patterns with a broader-scale circulation system or climate mode (Bunn, 2008).

3 Results and discussion

3.1 Spatio-temporal distribution of tree-ring chronology and hydrological station networks in Nepal

We developed a network of 106 tree-ring chronologies using 12 annual tree-ring-producing tree taxa distributed in an elevation range from 1220 to 4100 m (Fig. 1). The average starting calendar date of the chronology was AD 1715 (median = 1733), and the average end date of the chronology was AD 2005 (median = 2011) (Fig. 2). The data of river gauge stations (103 stations) of Nepal cover from 1962 to 2020 (Figs. 1 and 2). Some stations are already closed, as indicated by Fig. 2. Some cover data for only a few years.

3.2 Hydrological sensitivity or potential of tree-ring chronologies

Our chronology network incorporates dendro-climatically suitable species and can contribute to extending the short hydrological records. The hydrological responses of tree-ring sites or regional composite chronologies revealed significant correlations with the monthly discharge of different months and seasons (Fig. 3). Our chronology showed significant responses with the monthly and seasonal streamflow at more than 30 river gauge stations; however, the response of many of the rivers is not strong enough to proceed with the reconstruction. Most of the tree-ring chronologies from the eastern and central parts of Nepal are relatively weakly correlated

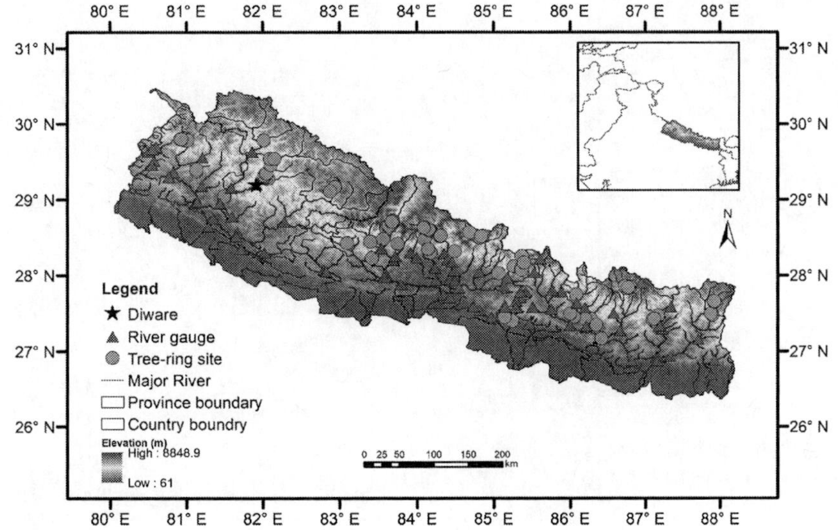

Figure 1. Spatial distribution of tree-ring width chronology (circular symbols) and river gauge station (triangles) networks in Nepal and Diware gauge station (star) in Jumla. Publisher's remark: please note that the above figure contains disputed territories.

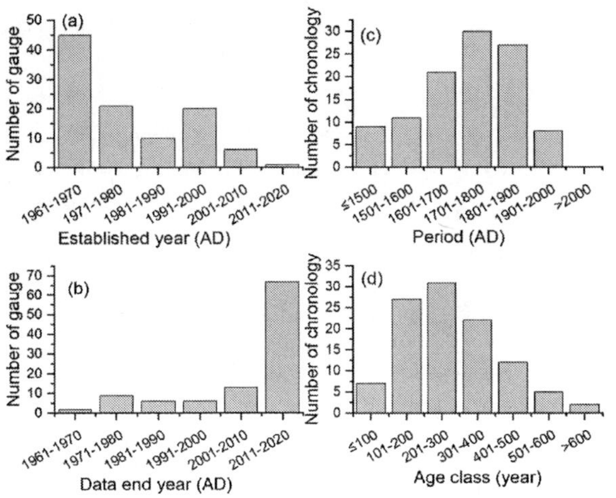

Figure 2. Temporal coverage of number of river gauges **(a, b)** and tree-ring chronology in Nepal **(c, d)**.

with the streamflow. However, several tree-ring chronologies from the western region revealed strong hydrological signals (Fig. 3).

A composite tree-ring width chronology named the "Sinja chronology" was prepared and used for reconstruction (Fig. 4a). Tree-ring chronology statistics indicate that the Sinja chronology meets the commonly applied criteria of EPS and Rbar statistics and shows their potential for possible hydro-climatic reconstruction (Fritts, 1976; Speer, 2010).

Hydrological response analysis of tree growth revealed a strong positive relationship between the Sinja chronology with monthly streamflow during March to July and the seasonal averages of the spring (MAM), summer (JJAS), and spring–early-summer (MAMJJ) seasons (Fig. 4b) of the current year. The tree-ring chronologies from the Sinja region generally revealed a positive relationship with spring-season (in some cases, both spring- and summer-season) precipitation and drought indices and negative correlations with the temperature during the spring season and, in some cases, also during the summer season (graph not shown). This indicates that the growth of the tree species in the studied region is mainly limited by moisture stress during the beginning or entirety of the growing season.

The streamflow is the result of rainfall, meltwater from snow and glaciers, and water from groundwater storage. The positive relations with streamflow (in addition to the positive response to precipitation) further confirm that the moisture availability limited tree growth in the Sinja valley and adjacent areas. Western Nepal (Karnali basin) receives about 7 % and 12 % of annual precipitation during the winter and spring seasons, respectively (Khatiwada et al., 2016). During the spring season, temperatures generally increase fast and reach their annual maximum. If there is less rainfall during winter and spring, an increasing demand for evaporation due to high temperatures in the late spring leads to moisture stress, limiting the growth of trees (Thapa et al., 2015; Panthi et al., 2017; Liang et al., 2019; Misra et al., 2015).

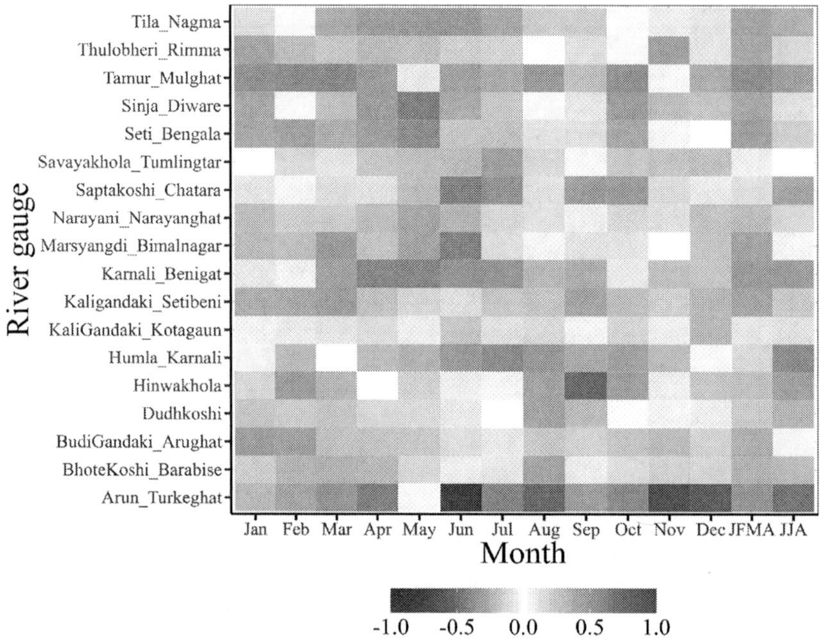

Figure 3. Hydro-climatic potential of the tree-ring chronologies from Nepal based on the relationship (correlation coefficient) in terms of tree-ring width between regional or site-based chronology and streamflow data.

3.3 Seasonal streamflow reconstruction of the Sinja River

The spring–early-summer season (March–July, MAMJJ) streamflow reconstructions of the Sinja River at Diware, Jumla, using a robust model ($\mathrm{MAMJJ_{Streamflow}} = 21.062 \times$ Tree ring index $- 1.3524$) that explains 56 % of the variance in the actual data (Fig. 4c) cover the periods AD 1700 to 2013 and show several pluvial (wet) and drought (dry) periods, with a long decreasing pattern since about 1890 to the 1950s until after the 1980s (Fig. 4d). The low-frequency analysis using a 10-year smoothing spline revealed several persistent dry and wet periods over the reconstruction period (Fig. 4d). Long above-average flow periods (calculated using 10-year smoothing-spline data) were observed during 1726–1736, 1752–1762, 1796–1803, 1826–1842, 1878–1906, 1911–1919, 1924–1934, 1973–1991, and 2007–2013, while long below-average flow periods occurred during 1703–1716, 1737–1751, 1763–1772, 1776–1787, 1804–1813, 1869–1877, 1942–1961, 1964–1972, 1992–1999, and 2002–2006. The recent dry and pluvial periods are synchronous to the dry and wet periods in the precipitation and drought reconstructions from the region (Gaire et al., 2017, 2019; Panthi et al., 2017). Several consecutive moderate to extremely dry and pluvial years (flow beyond 1 and 2 SD from the long-term average) were also recorded. Some of the extremely dry years found in the present study match with the station-data-based drought years (1984–1985, 1987–1988, 1992–1993, 1994–1995, 1999, 2004–2009, and 2012) in the Karnali River basin (Khatiwada and Pandey, 2019). Many extreme-flow periods were observed in the lower reaches of the Karnali basin (Gaire et al., 2022).

Centennial to millennial long tree-ring-based seasonal to annual streamflow reconstructions have been carried out for several rivers in the western Himalayas (Cook et al., 2013; Singh and Yadav, 2013; Mishra et al., 2015) and eastern Himalayas (Shah et al., 2014; Rao et al., 2020), as well as for many rivers in China (Chen et al., 2019; Liu et al., 2020). Some of the dry and wet events of our study match with those streamflow reconstructions from eastern Himalayan rivers like the Lachen "Chu" River, the North Sikkim (Shah et al., 2014), and the Brahmaputra River (Rao et al., 2020), though rivers in the eastern Himalayas are more influenced by the Indian summer monsoon (ISM). The climate of Nepal's (central) Himalayas is influenced by both the ISM and western disturbances; some of the hydro-climatic events match with studies of the western Himalayas, and some match with studies of the eastern Himalayas. A decreasing tendency in Satluj River discharge (Misra et al., 2015), reconstructed using tree-ring data, was recorded from the 1990s, which is similar to our reconstructions.

We observed interannual- to multidecadal-scale (medium-term) periodicities in our March–July streamflow in the Sinja River, as revealed by the power spectral analysis (Fig. 5). Significant (95 % or above) spectral peaks were observed in the periods of 2.2–2.9, 6.2, 6.3, 8.2, 8.7, 35.2, and

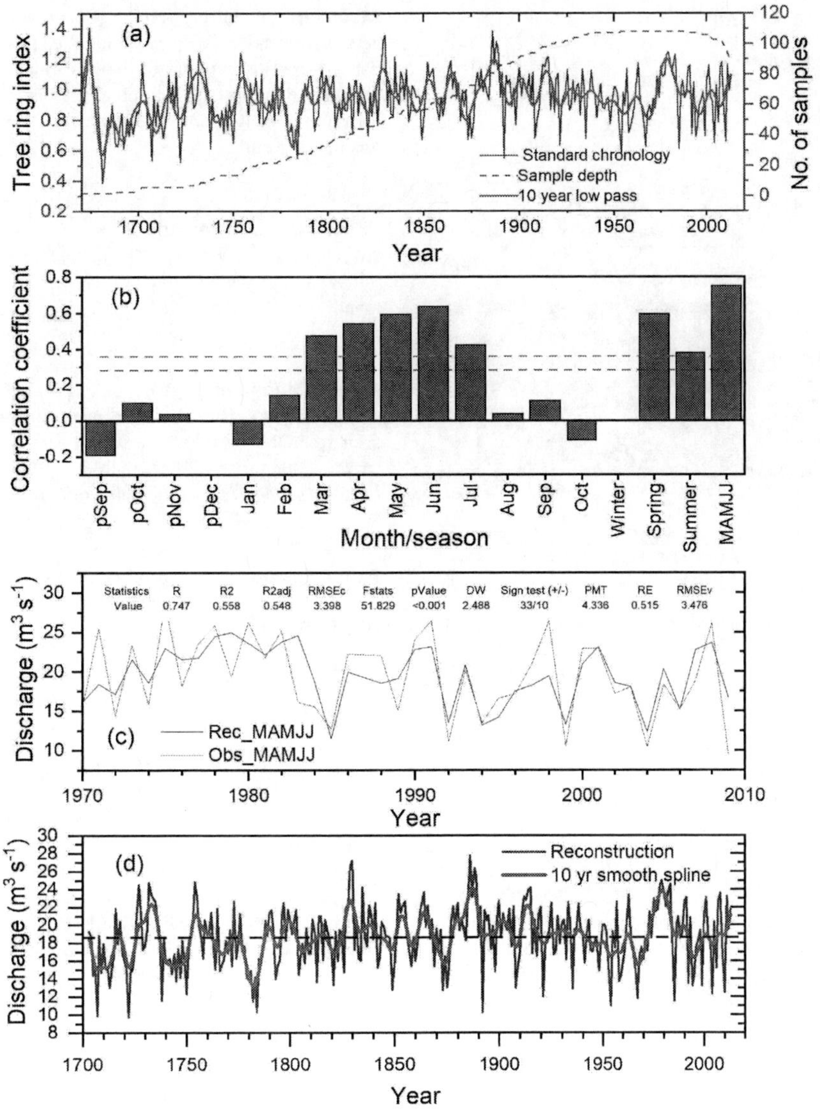

Figure 4. Composite tree-ring standard chronology from Sinja River basin, along with the 10-year low-pass smoothing and sample depth **(a)** and the hydrological response, i.e. relationship (Pearson's correlation coefficient) between the composite tree-ring standard chronology and the monthly and seasonal streamflow data **(b)**. MAMJJ indicates spring–early-summer season. The horizontal dashed line indicates 95 % (black) and 99 % (red) levels of significance of the correlation coefficients. Calibration models **(c)** and reconstructed spring–summer season (March–July) average streamflow **(d)** in the Sinja Khola, Karnali basin.

36.5 years (Fig. 5). The short and interannual periodicities of 2 to 8 years fall within the El Niño–Southern Oscillation (ENSO) range, which has quasi-biennial variability with a 2.1–2.9-year frequency, together with a lower-frequency (2.5–8.7 years) component (Bridgman and Oliver, 2006). The medium-term (35.2–36.5 years) periodicities are likely to be related to the influence of the Atlantic Multidecadal Oscillation (AMO) and the North Atlantic Oscillation (NAO). The periodicity related to ENSO is one of the dominant forcings in relation to local dryness–wetness variations in the South Asian summer-monsoon-dominated Himalayas (Yadav et al., 2014; Gaire et al., 2017, 2019; Panthi et al., 2017), as well as in the Tibetan plateau and adjacent regions (Gou et al., 2010; Chen et al., 2019).

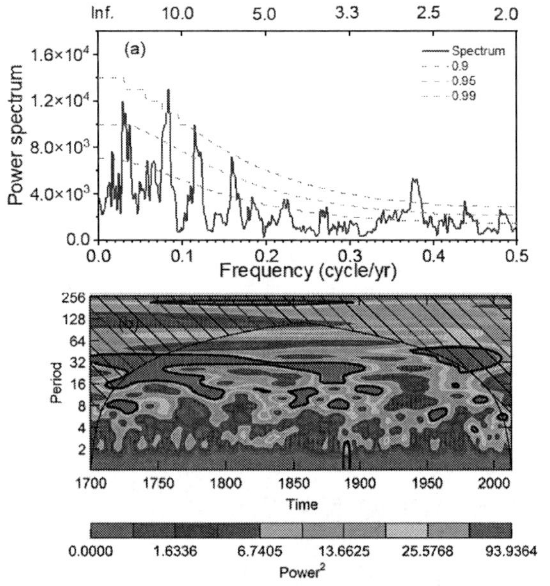

Figure 5. Multi-tapered power spectral analysis **(a)** and wavelet analysis **(b)** of reconstructed streamflow of the Sinja Khola (River) at Diware, Karnali basin.

4 Conclusions

We present the hydrological sensitivity of multiple tree species from the Nepalese Himalayas and their potential to be used for long-term streamflow reconstructions. The over-3-centuries-long March–July streamflow reconstructions of the Sinja Khola in the Nepalese Himalayas revealed periodic fluctuations in the streamflow, with a continuous decrease after the 1980s, indicating possible water shortages in the warming but drying world anticipated in the future. Since we found periodic (decadal or even longer) below-average streamflows during the last 3 centuries, the streamflow-based water and energy sectors could be affected periodically.

Data availability. Hydrological data were obtained from the Department of Hydrology and Meteorology, Government of Nepal. Public tree-ring data were accessed from NOAA. The reconstructed data will be available on request to the authors.

Author contributions. NPG did the conceptualization, methodology, formal analysis, and writing – original draft. SKS and FZX also participated in the conceptualization, methodology, and writing – review and editing. YRD did the writing – review and editing.

Competing interests. The contact author has declared that none of the authors has any competing interests.

Disclaimer. Publisher's note: Copernicus Publications remains neutral with regard to jurisdictional claims made in the text, published maps, institutional affiliations, or any other geographical representation in this paper. While Copernicus Publications makes every effort to include appropriate place names, the final responsibility lies with the authors.

Special issue statement. This article is part of the special issue "Mountain Hydrology and Cryosphere". It is a result of the International Conference on Mountain Hydrology and Cryosphere, Kathmandu and Dhulikhel, Nepal, 9–10 November 2023.

Acknowledgements. We thank DHM for the hydro-meteorological data, the Tribhuvan University for the research grant, the DNPWC for the research permissions, NAST and FRTC for the laboratory facilities, and the local field assistants and guides for their support during the fieldwork.

Financial support. This research has been supported by the Tribhuvan University (grant nos. TU-NPAR-079/80-ERG12 and TU-HERP DLI-7B).

References

Bolch, T., Kulkarni, A., Kääb, A., Huggel, C., Paul, F., Cogley, J. G., Frey, H., Kargel, J. S., Fujita, K., Scheel, M., Bajracharya, S., and Stoffel, M.: The state and fate of Himalayan glaciers, Science, 336, 310–314, https://doi.org/10.1126/science.1215828, 2012.

Bridgman, H. A. and Oliver, J. E.: The global climate system: patterns, processes, and teleconnections, Cambridge University Press, Cambridge, ISBN 9781107668379, 2006.

Bunn, A. G.: A dendrochronology program library in R (dplR), Dendrochronologia, 26, 115–124, https://doi.org/10.1016/j.dendro.2008.01.002, 2008.

Chaudhary, S., Chettri, N., Adhikari, B., Dan, Z., Gaire, N. P., Shrestha, F., and Wang, L.: Effects of a changing cryosphere on biodiversity and ecosystem services, and response options in the Hindu Kush Himalaya, in: ICIMOD, Water, ice, society, and ecosystems in the Hindu Kush Himalaya: An outlook, edited by: Wester, P., Chaudhary, S., Chettri, N., Jackson, M., Maharjan, A., Nepal, S., and Steiner, J. F., ICIMOD, 123–163, https://doi.org/10.53055/ICIMOD.1032, 2023.

Chen, F., Shang, H. M., and Panyushkina, I. P.: Tree-ring reconstruction of Lhasa River streamflow reveals 472 years of hydrologic change on southern Tibetan Plateau, J. Hydrol., 572, 169–178, https://doi.org/10.1016/j.jhydrol.2019.02.054, 2019.

Cook, E. R. and Kairiukstis, A. (Eds.): Methods of Dendrochronology: Applications in the Environmental Sciences, Kluwer Academic Press, Dordrecht, ISBN 13:978-0792305866, ISBN 10:0792305868, 1990.

Cook, E. R., Palmer, J. G., Ahmed, M., Woodhouse, C. A., Fenwick, P., Zafar, M. U., Wahab, M., and Khan, N.: Five centuries of Upper Indus River flow from tree rings, J. Hydrol., 486, 365–375, https://doi.org/10.1016/j.jhydrol.2013.02.004, 2013.

DHM: Observed Climate Trend Analysis in the Districts and Physiographic Regions of Nepal (1971–2014), Department of Hydrology and Meteorology, Kathmandu, https://www.dhm.gov.np/uploads/dhm/climateService/Observed_Climate_Trend_Analysis_Report_2017.pdf (last access: 28 July 2024), 2017.

Fritts, H. C.: Tree rings and climate, Academic Press, New York, ISBN 9781930665392, 1976.

Gaire, N., Zaw, Z., Bräuning, A., Sharma, B., Dhakal, Y., Timilsena, R., Shah, S. K., Bhuju, D. R., and Fan, Z.: Increasing extreme events in the central Himalaya revealed from a tree-ring based multi-century streamflow reconstruction of Karnali River Basin, J. Hydrol., 610, 127801, https://doi.org/10.1016/j.jhydrol.2022.127801, 2022.

Gaire, N. P., Koirala, M., Bhuju, D. R., Shah, S. K., Carrer, M., and Timilsena, R. R.: Tree-ring based spring precipitation reconstruction in western Nepal Himalaya since AD 1840, Dendrochronologia, 42, 21–30, https://doi.org/10.1016/j.dendro.2016.12.004, 2017.

Gaire, N. P., Dhakal Y. R., Shah, S. K., Fan, Z. X., Bräuning A., Thapa, U. K., Bhandari, U. K., Aryal, S., and Bhuju, D. R.: Drought (scPDSI) reconstruction of trans-Himalayan region of central Himalaya using Pinus wallichiana tree-rings, Palaeogeogr. Palaeoclim. Palaeoecol., 514, 251–264, https://doi.org/10.1016/j.palaeo.2018.10.026, 2019.

Gaire, N. P., Fan, Z.-X., Chhetri, P. K., Shah, S. K., Bhuju, D. R., Wang, J., Sharma, B., Shi, P., and Dhakal, Y. R.: Treeline Dynamics in Nepal Himalaya in a Response to Complexity of Factors, in: Ecology of Himalayan Treeline Ecotone, edtied by: Singh, S. P., Reshi, Z. A., and Joshi, R., Springer, Singapore, 519–563, https://doi.org/10.1007/978-981-19-4476-5_22, 2023.

Gou, X. H., Deng, Y., Chen, F., Yang, M., Fang, K., Gao, L., Yang, T., and Zhang, F.: Tree ring-based streamflow reconstruction for the Upper Yellow River over the past 1234 years, Chin. Sci. Bull., 55, 4179–4186, https://doi.org/10.1007/s11434-010-4215-z, 2010.

ICIMOD: Water, ice, society, and ecosystems in the Hindu Kush Himalaya: An outlook, edited by: Wester, P., Chaudhary, S., Chettri, N., Jackson, M., Maharjan, A., Nepal, S., and Steiner, J. F., ICIMOD, https://doi.org/10.53055/ICIMOD.1028, 2023.

Immerzeel, W. W., van Beek, L. P. H., and Bierkens, M. F. P.: Climate change will affect the Asian water towers, Science, 328, 1382–1385, https://doi.org/10.1126/science.1183188, 2010.

Khatiwada, K. R. and Pandey, V. P.: Characterization of hydro-meteorological drought in Nepal Himalaya: A case of Karnali River Basin, Weather Clim. Extrem., 26, 100239, https://doi.org/10.1016/j.wace.2019.100239, 2019.

Khatiwada, K., Panthi, J., Shrestha, M., and Nepal, S.: Hydro-climatic variability in the Karnali River basin of Nepal Himalaya, Climate, 6, 17, https://doi.org/10.3390/cli4020017, 2016.

Liang, E. Y., Dawadi, B., Pederson, N., Piao, S., Zhu, H., Sigdel, S. R., and Chen, D.: Strong link between large tropical volcanic eruptions and severe droughts prior to monsoon in the central Himalayas revealed by tree-ring records, Sci. Bull., 64, 1018–1023, https://doi.org/10.1016/j.scib.2019.05.002, 2019.

Liu, Y., Song, H. M., and An, Z. S.: Recent anthropogenic curtailing of yellow river runoff and sediment load is unprecedented over the past 500 y, P. Natl. Acad. Sci. USA, 117, 18251–18257, https://doi.org/10.1073/pnas.1922349117, 2020.

Mathison, C., Wiltshire, A. J., Falloon, P., and Challinor, A. J.: South Asia river-flow projections and their implications for water resources, Hydrol. Earth Syst. Sci., 19, 4783–4810, https://doi.org/10.5194/hess-19-4783-2015, 2015.

Michaelsen, J.: Cross-validation in statistical climate forecast models, J. Clim. Appl. Meteorol., 26, 1589–1600, 1987.

Misra, K. G., Yadav, R. R., and Misra, S.: Satluj river flow variations since AD 1660 based on tree-ring network of Himalayan cedar from western Himalaya, India, Quatern. Int., 371, 135–143, https://doi.org/10.1016/j.quaint.2015.01.015, 2015.

Panthi, S., Bräuning, A., Zhou, Z. K., and Fan, Z. X.: Tree rings reveal recent intensified spring drought in the central Himalaya, Nepal, Global Planet. Change 157, 26–34, https://doi.org/10.1016/j.gloplacha.2017.08.012, 2017.

Rao, M. P., Cook, E. R., and Cook, B. I.: Seven centuries of reconstructed Brahmaputra River discharge demonstrate underestimated high discharge and flood hazard frequency, Nat. Commun., 11, 1–10, https://doi.org/10.1038/s41467-020-19795-6, 2020.

Rinn, F.: TSAP-Win: time series analysis and presentation for dendrochronology and related applications, Version 0.55 User reference, Heidelberg, Germany, https://rinntech.info/products/tsap-win/ (last access: 28 July 2024), 2013.

Shah, S. K., Bhattacharyya, A., and Chaudhary, V.: Streamflow reconstruction of Eastern Himalaya River, Lachen 'Chhu', North Sikkim, based on tree-ring data of Larix griffithiana from Zemu Glacier basin, Dendrochronologia, 32, 97–106, https://doi.org/10.1016/j.dendro.2014.01.005, 2014.

Singh, J. and Yadav, R. R.: Tree ring based seven century long flow record of Satluj River western Himalaya, India, Quatern. Int., 304, 156–162, https://doi.org/10.1016/j.quaint.2013.03.024, 2013.

Speer, J. H.: Fundamentals of tree ring research, The University of Arizona Press, Tucson, ISBN 13:978-0816526857, ISBN 10:0816526850, 2010.

Talchabhadel, R., Karki, R., Thapa, B. R., Maharjan, M., and Parajuli, B.: Spatio-temporal variability of extreme precipitation in Nepal, Int. J. Climatol., 38, 4296–4313, https://doi.org/10.1002/joc.5669, 2018.

Thapa, U. K., Shah, S. K., Gaire, N. P., and Bhuju, D. R.: Spring temperatures in the far-western Nepal Himalaya since AD 1640 reconstructed from Picea smithiana tree-ring widths, Clim. Dynam., 45, 2069–2081, https://doi.org/10.1007/s00382-014-2457-1, 2015.

Wigley, T. M. L., Briffa, K. R., and Jones, P. D.: On the average value of correlated time series, with applications in dendroclimatology and hydrometeorology, J. Clim. Appl. Meteorol., 23, 201–213, https://doi.org/10.1175/1520-0450(1984)023<0201:OTAVOC>2.0.CO;2, 1984.

Yadav, R. R., Misra, K. G., Kotlia, B. S., and Upreti, N.: Premonsoon precipitation variability in Kumaon Himalaya, India over a perspective of ∼ 300 years, Quatern. Int., 325, 213–219, https://doi.org/10.1016/j.quaint.2013.09.005, 2014.

Zang, C. and Biondi, F.: Treeclim: an R package for the numerical calibration of proxy-climate relationships, Ecography, 38, 431–436, https://doi.org/10.1111/ecog.01335, 2015.

Proc. IAHS, 387, 41–46, 2024
https://doi.org/10.5194/piahs-387-41-2024

Open Access PIAHS

Using radiotracers ^{137}Cs and ^{210}Pb$_{ex}$ to document climate change in mountain areas through the estimate of soil erosion rates

Paolo Porto[1,2]

[1]Dipartimento di Agraria, Università degli Studi Mediterranea, Località Feo di Vito, Reggio Calabria, Italy
[2]Faculty of Geographical Sciences, Kazimierz Wielki University, 85-033 Bydgoszcz, Poland

Correspondence: Paolo Porto (paolo.porto@unirc.it)

Published: 18 November 2024

Abstract. During the last few decades, a general increase in heavy rainfall events has been documented in many areas of the world. These events have caused changes in soil erosion rates and strongly affected the human activities in mountain areas by reducing the national income obtained from cultivated land. In this context, the use of fallout radiotracers can be an important tool for better understanding the consequences of climate change in these areas and proposing effective countermeasures in order to reduce soil loss. In this contribution, plot experiments carried out in southern Italy that involve the use of ^{137}Cs and ^{210}Pb$_{ex}$ measurements were performed to estimate soil erosion rates during the last few decades. The overall results indicate an increase in soil erosion rates during the last 15–20 years and suggest the use of this technique to detect climate change in mountain areas.

1 Introduction

The global fallout of the artificial radionuclide caesium-137 (^{137}Cs; half-life of ca. 30.2 years), associated with the testing of nuclear weapons performed during the period 1954–1980, has now been successfully used as the basis for estimating soil erosion rates in many areas of the world (IAEA, 2014; Ritchie and Ritchie, 2005). Important advantages of the approach include, firstly, the potential to obtain retrospective information on soil erosion rates on the basis of a single site visit; secondly, the spatially distributed nature of the data generated; and, thirdly, the ability to collect data from the "natural" landscape without the need to establish long-term monitoring using experimental plots or catchments that have been proven to be very expensive and time-consuming. In nearly all situations, the resulting estimates of the soil erosion rate represent values of mean annual erosion for a time window representing the period elapsed between the onset of significant ^{137}Cs fallout in the mid-1950s (or the occurrence of peak ^{137}Cs fallout in the early 1960s) and the time of sample collection.

The similar behaviour of fallout lead-210 (^{210}Pb; half-life ca. 22 years) in soils makes it a potential alternative to ^{137}Cs for soil erosion investigations in areas where ^{137}Cs measurements encounter problems. In fact, thanks to its continuous fallout (depending only on precipitation); its estimates cover a longer period (ca. 110 years = 5 half-lives); and, therefore, its combined use with ^{137}Cs can provide important suggestions in an extended time window. The potential to obtain estimates of soil erosion rates related to different time windows can also be exploited to document possible recent changes in soil loss due to variation in land use and/or due to climate change. In fact, by virtue of its shorter half-life, ^{210}Pb$_{ex}$ results more sensitive to soil erosion rates occurred during the last 15–20 years, and the problem of documenting changing erosion rates over such period can be addressed.

This paper describes this technique based on the use of ^{137}Cs and ^{210}Pb$_{ex}$ measurements collected within some experimental plots located in two mountain areas of southern Italy.

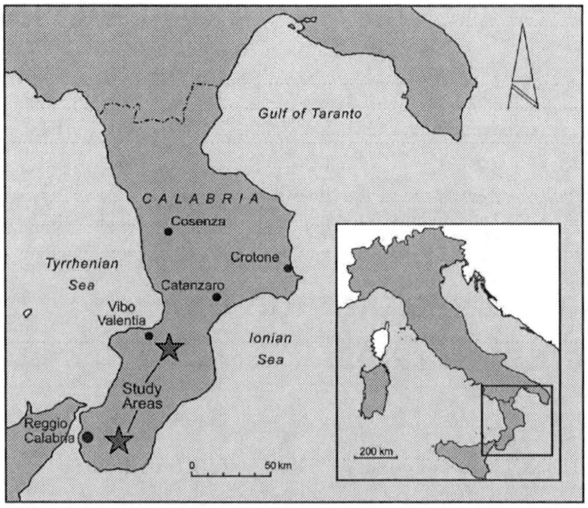

Figure 1. The location of the two study areas in southern Italy.

2 The study areas

The investigation was carried out in two study areas (Mongiana and Ferraina) located across the Calabrian Apennine mountain and both covered by beech (*Fagus sylvatica*) and pine (*Pinus nigra laricio*) forests (Fig. 1). Both areas experience a temperate climate, with mild to warm summers and cool winters.

In the experimental site of Mongiana (range of elevation = 850–1050 m a.s.l.), three areas (3 ha in size) have been identified. In each area, three plots have been established to document soil erosion rates in three sub-areas in response to different silvicultural thinning practised during the last 30 years. The sub-areas consist of an undisturbed zone (control), here selected as a reference site; a zone with a traditional thinning (namely Plot 1), in which a cutting with moderate intensity was operated; and a zone with an innovative treatment (namely Plot 2), in which ca. 27 % of the total biomass was removed.

In the experimental site of Ferraina (range of elevation = 1360–1465 m a.s.l.), four plots, including two stands with Calabrian pine of different canopy covers and two stands with beech (one of which was cut in 2009), have been established. More specifically, the two pine forest stands, namely Plot 1 and Plot 2, are characterized by 50-year pine trees planted in the late 1960s during a massive re-afforestation programme (ca. 150 000 ha). The two beech forest stands, namely Plot 3 (beech forest clear-cut) and Plot 4 (beech forest undisturbed) consisted of seminatural high beech forests (ca. 140 years old). Plot 4 was not disturbed by cutting activities, except for a minimum removal of trees in the past few decades; on the contrary, Plot 3 was clear-cut in 2009 and then left to evolve naturally.

3 The sampling and analysis programme for ^{137}Cs measurements

The sampling programme for ^{137}Cs and ^{210}Pb$_{ex}$ measurements comprised several field campaigns aimed at collecting soil samples to characterize an undisturbed location (reference site) and the plots themselves.

The sampling was undertaken in May 2018 in the experimental area of Mongiana and in June 2018 in Ferraina using the sampling strategy illustrated in Fig. 2.

Totally, eight final composite soil cores (six from the plots and two from the reference areas) were taken with a 10 cm diameter steel core tube driven into the soil to a depth of ca. 35 cm. Each composite core taken from the plots consisted of three single-sectioned cores combined layer by layer. Each sectioned core consisted of ca. 15–20 layers of 2 cm each. In the site of Ferraina, six additional composite bulk samples were collected for each plot to account for spatial variability. The composite core collected in the two reference areas consisted of nine single-sectioned cores combined layer by layer.

All samples provided by the sampling campaigns, namely the composite bulk cores and the sectioned cores, as well as the samples collected from the reference site were initially oven-dried at 105 °C, disaggregated, homogenized and passed through a 2 mm sieve. A representative fraction of the <2 mm fraction of each sample was then placed into a 0.5 L Marinelli beaker or, in the case of smaller samples, plastic pots (330 cm^3) for determination of their ^{137}Cs and ^{210}Pb$_{ex}$ activity. Before the analyses, the samples were sealed for 3 weeks based on the standard requirements for ^{210}Pb$_{ex}$ measurements. The samples were analysed at the Department of Agriculture of the Mediterranea University in Reggio Calabria, Italy, using high-resolution high-purity germanium (HPGe) detectors. The total inventory (Bq m^{-2}) of each bulk core (see Table 1) was calculated as the product of the measured ^{137}Cs activity (Bq kg^{-1}) and the dry mass of the <2 mm fraction of the bulk core (kg), divided by the surface area associated with the core (m^2).

4 Results

4.1 ^{137}Cs and ^{210}Pb$_{ex}$ inventories at the reference sites and within the study plots

Information regarding the depth distribution of ^{137}Cs and ^{210}Pb$_{ex}$ at the reference sites is reported in Fig. 3 for both study areas.

The total inventory associated with the ^{137}Cs profiles is estimated to be 5949 Bq m^{-2} for Ferraina and 4697 Bq m^{-2} for Mongiana (see Table 1). For the ^{210}Pb$_{ex}$ profiles, the total inventory is estimated to be 10 168 Bq m^{-2} for Ferraina and 14 400 Bq m^{-2} for Mongiana. Table 1 provides a summary of the mean, the range and the standard deviation of

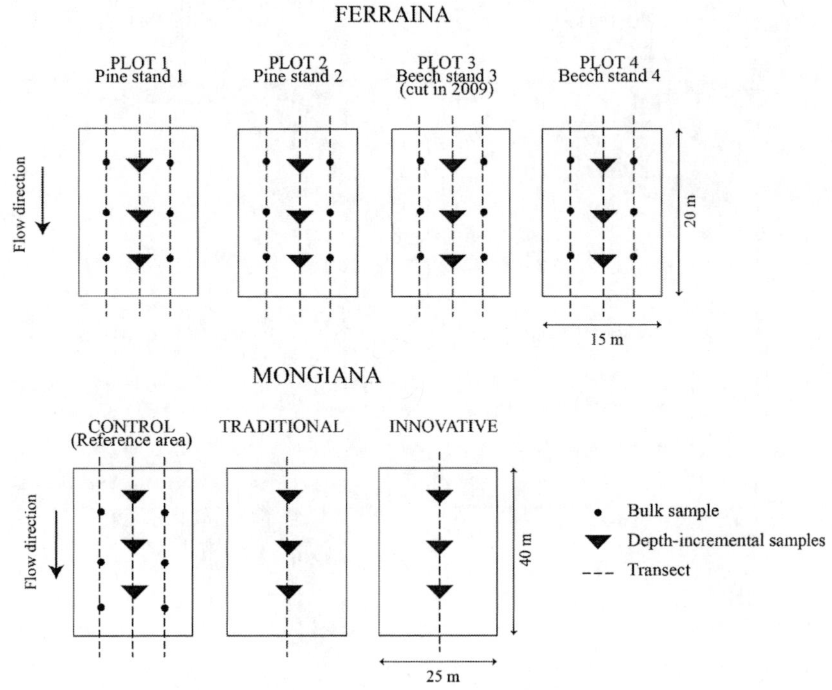

Figure 2. The sampling design within the experimental plots and in the reference areas.

Table 1. ^{137}Cs and ^{210}Pb$_{ex}$ inventories for the reference site and for the study plots.

	^{137}Cs				^{210}Pb$_{ex}$			
	Reference inventory	Inventory across the plots			Reference inventory	Inventory across the plots		
Study area	[Bq m^{-2}]	Mean [Bq m^{-2}]	Range [Bq m^{-2}]	SD [Bq m^{-2}]	[Bq m^{-2}]	Mean [Bq m^{-2}]	Range [Bq m^{-2}]	SD [Bq m^{-2}]
Ferraina	5949	4656	1374–10 943	2053	10 168	7985	902–16 683	4313
Mongiana	4697	3122	1490–4322	1205	14 400	6837	5020–9229	1784

the ^{137}Cs; the ^{210}Pb$_{ex}$ inventories obtained within the experimental plots; and the reference sites for the two study areas.

4.2 Estimating erosion rates from the ^{137}Cs and ^{210}Pb$_{ex}$ measurements obtained for the sampling points within the study plots

Estimation of erosion rates from ^{137}Cs and ^{210}Pb$_{ex}$ measurements is generally based on the degree of reduction in the measured inventory relative to the local reference inventory. In this study, a diffusion and migration model has been used for the purpose of estimating the soil erosion rates indicated by the ^{137}Cs and ^{210}Pb$_{ex}$ inventories measured at the sampling sites (see Porto et al., 2003, 2004, 2006, 2018). The model converts the magnitude of the reduction or increase in the radionuclide inventory relative to the reference inventory to an estimate of the rate of soil loss for uncultivated sites. Further details of this model can be found in Porto et al. (2003, 2018).

The range of the erosion rates estimated for the individual sampling points within the study plots based on the ^{137}Cs and ^{210}Pb$_{ex}$ measurements are presented in Fig. 4 for each study area. The dashed lines representing the mean value of these estimates are also superimposed in the graphs for further comparison.

5 Discussion

The results reported in Table 1 and Fig. 4 indicate that appreciable rates of soil erosion have occurred throughout the study plots during the time windows explored by the two radiotracers. These rates (see Fig. 4) are higher for the Mongiana site, where the values of soil loss provided by ^{137}Cs

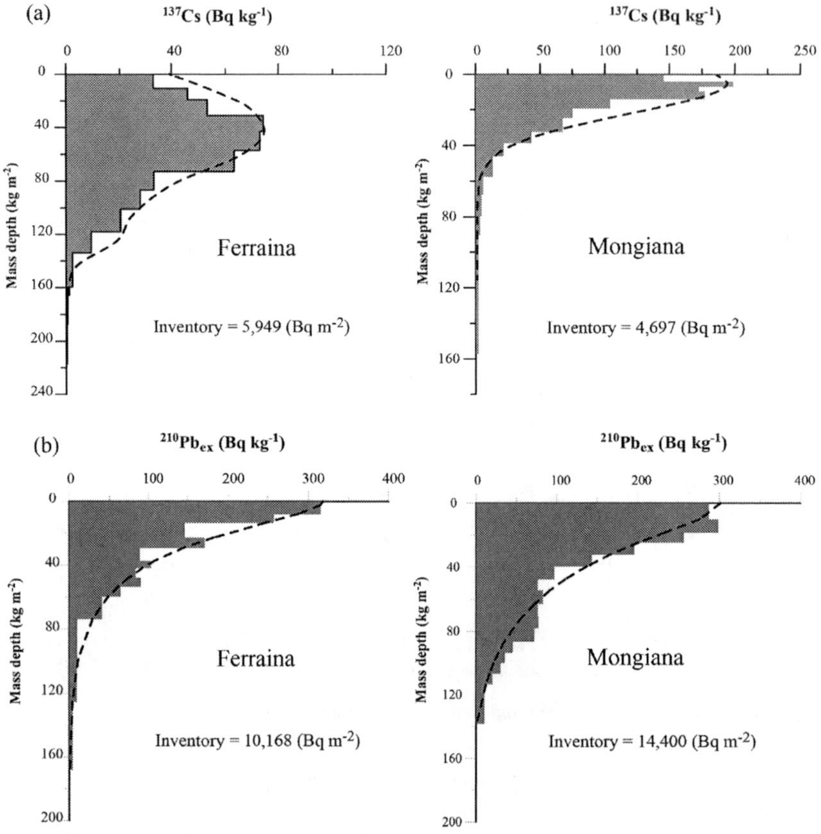

Figure 3. The ^{137}Cs **(a)** and the ^{210}Pb$_{ex}$ **(b)** depth distributions at the reference sites.

range from 2.55 to 9.85 t ha^{-1} yr^{-1}, and those obtained from ^{210}Pb$_{ex}$ measurements range from 6.45 to 10.75 t ha^{-1} yr^{-1}. Equivalent estimates at the Ferraina site provided lower values, ranging from 0.25 to 1.33 t ha^{-1} yr^{-1} for ^{137}Cs and from 0.77 to 3.50 t ha^{-1} yr^{-1} for ^{210}Pb$_{ex}$ measurements. This difference in magnitude between the two sites is related to a number of factors that include difference in rainfall erosivity, change in land use and silviculture treatments, as it was emphasized by Altieri et al. (2018) and Romeo et al. (2021). However, when the comparison is related to the two sets of estimates provided by ^{137}Cs and ^{210}Pb$_{ex}$ measurements within the same study area, it can be seen that overall values of erosion rates obtained with ^{210}Pb$_{ex}$ are higher than those provided by ^{137}Cs. These estimates of mean annual soil loss produced for the two study areas are also outlined in Fig. 4 (see dashed lines). The estimates of the long-term mean erosion rates derived from the ^{137}Cs measurements relate to a period of about 65 years, extending from the commencement of fallout (1954) to the time of sampling. In contrast, because ^{210}Pb$_{ex}$ fallout can be considered to be essentially continuous, the equivalent estimates derived from the ^{210}Pb$_{ex}$ measurements will reflect a longer period of up to 110 years (ca. 5 half-lives). However, because of the relatively short half-life of ^{210}Pb$_{ex}$ (22.3 years), the final estimates of erosion rates obtained using this radiotracer are likely to be more sensitive to erosion occurring during the past 15–20 years. The higher soil erosion rates estimated using the ^{210}Pb$_{ex}$ measurements are therefore seen as a reflection of increased erosional activity in recent years in response to, for example, changes in rainfall patterns or, alternatively, changes in land use and/or silvicultural treatment (Porto and Walling, 2012; Porto et al., 2009). The increased soil loss evidenced by the experimental plots is consistent with the results of recent studies undertaken in southern Italy, where a trend of increasing rainfall erosivity during the last 15–20 years has been documented (Porto and Walling, 2012; Porto et al., 2013; Capra et al., 2017). In Fig. 5, the annual values of the rainfall erosivity factor (R factor) (Wischmeier and Smith, 1978) for the two stations of Serra San Bruno, located in the Mongiana study area, and of Gambarie d'Aspromonte, located in the vicinity of the Ferraina site, are reported. The calculations of the R factor were made using rainfall datasets related to a duration of 30 min (see Porto, 2016). The histograms presented in Fig. 5 confirm the increasing trend of the annual values of

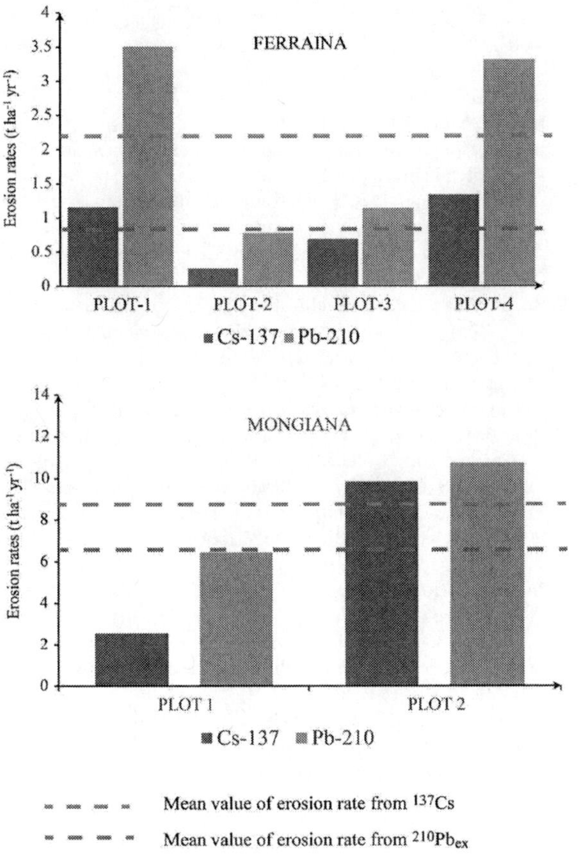

Figure 4. Comparison between the ranges covered by the estimates of mean annual soil loss provided by ^{137}Cs and ^{210}Pb$_{ex}$ for each study area.

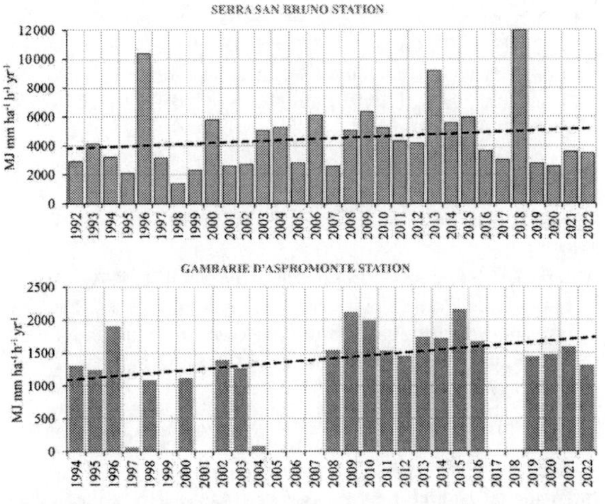

Figure 5. Annual values and trends of the rainfall erosivity factor calculated for the Serra San Bruno station, located in the Mongiana study area, and for the Gambarie d'Aspromonte station, located in the vicinity of the Ferraina site.

the rainfall erosivity factor during the last 30 years for both study areas.

A comparison between the two sets of data also reveals a higher magnitude of rainfall erosivity for the Mongiana site than for Ferraina, where the values of the R factor are ca. 1 : 5 of those of the former. This difference also confirms the higher absolute estimates of soil erosion rates provided by the two radiotracers for the two study areas.

6 Conclusions

The results presented above are seen as a confirmation of the potential for combining ^{137}Cs and ^{210}Pb$_{ex}$ measurements to document recent changes in soil erosion in mountain areas. The results obtained from the ^{210}Pb$_{ex}$ measurements point to higher values of soil loss if compared to equivalent estimates provided by ^{137}Cs. This is consistent with the trend demonstrated by independent information on changes in annual rainfall erosivity in these areas during the last 15–20 years. Further validation of this approach, particularly in different environments and in areas where soil erosion rates might be expected to be declining, could usefully be undertaken.

Data availability. Data are available upon request from the author.

Competing interests. The author has declared that there are no competing interests.

Disclaimer. Publisher's note: Copernicus Publications remains neutral with regard to jurisdictional claims made in the text, published maps, institutional affiliations, or any other geographical representation in this paper. While Copernicus Publications makes every effort to include appropriate place names, the final responsibility lies with the authors.

Special issue statement. This article is part of the special issue "Mountain Hydrology and Cryosphere". It is a result of the International Conference on Mountain Hydrology and Cryosphere, Kathmandu and Dhulikhel, Nepal, 9–10 November 2023.

Acknowledgements. The study has been finalized in the frame of Erasmus+ KA2 – Cooperation for innovation and the exchange of good practices – Capacity building in the field of higher universities of western Balkan countries/SETOF. The author is also indebted to the Agenzia Regionale per la Protezione dell'Ambiente (ARPACAL) for providing the rainfall data used in this study.

Financial support. This research has been supported by the Erasmus+CBHE project titled "Soil Erosion and Torrential Flood Prevention: Curriculum Development at the Universities of Western Balkan Countries – SETOF" (grant no. 598403-EPP-1-2018-1-RS-EPPKA2-CBHE-JP).

References

Altieri, V., De Franco, S., Lombardi, F., Marziliano, P. A., Menguzzato, G., and Porto, P.: The role of silvicultural systems and forest types in preventing soil erosion processes in mountain forests. A methodological approach using Caesium-137 measurements, J. Soil. Sediment., 18, 3378–3387, 2018.

Capra, A., Porto, P., and La Spada, C.: Long-term variation of rainfall erosivity in Calabria (southern Italy), Theor. Appl. Climatol., 128, 141–158, 2017.

International Atomic Energy Agency (IAEA): Guidelines for Using Fallout Radionuclides to Assess Erosion and Effectiveness of Soil Conservation Strategies, IAEA-TECDOC-1741, IAEA, Vienna, https://www-pub.iaea.org/MTCD/Publications/PDF/TE-1741_web.pdf (last access: 19 July 2024), 2014.

Porto, P.: Exploring the effect of different time resolutions to calculate the rainfall erosivity factor R in Calabria, southern Italy, Hydrol. Process., 30, 1551–1562, 2016.

Porto, P. and Walling, D. E.: Validating the use of ^{137}Cs and ^{210}Pb$_{ex}$ measurements to estimate rates of soil loss from cultivated land in Southern Italy, J. Environ. Radioactiv., 106, 47–57, 2012.

Porto, P., Walling, D. E., Ferro, V., and Di Stefano, C.: Validating erosion rate estimates by caesium-137 measurements for two small forested catchments in Calabria, Southern Italy, Land Degrad. Dev., 14, 389–408, 2003.

Porto, P., Walling, D. E., and Callegari, G.: Validating the use of caesium-137 measurements to estimate erosion rates in three small catchments in Southern Italy, IAHS/AISH Publ., 288, 75–83, 2004.

Porto, P., Walling, D. E., Callegari, G., and Catona F.: Using fallout lead-210 measurements to estimate soil erosion in three small catchments in Southern Italy, Water, Air, And Soil Pollution: Focus, 6, 657–667, 2006.

Porto, P., Walling, D. E., and Callegari, G.: Investigating the effects of afforestation on soil erosion and sediment mobilisation in two small catchments in Southern Italy, Catena, 79, 181–188, 2009.

Porto, P., Walling, D. E., and Callegari, G.: Using ^{137}Cs and ^{210}Pb$_{ex}$ measurements to investigate the sediment budget of a small forested catchment in southern Italy, Hydrol. Process., 27, 795–806, 2013.

Porto, P., Walling, D. E., and Callegari ,G.: Using repeated ^{137}Cs and ^{210}Pb$_{ex}$ measurements to establish sediment budgets for different time windows and explore the effect of connectivity on soil erosion rates in a small experimental catchment in Southern Italy, Land Degrad. Dev., 29, 1819–1832, 2018.

Ritchie, J. C. and Ritchie, C. A.: Bibliography of publications of ^{137}Cesium studies related to erosion and sediment deposition, USDA-ARS Hydrology and Remote Sensing Lab, Occasional Paper HRSL-2005-01, https://www.ars.usda.gov/ARSUserFiles/80420510/Cesium137/BiblioCs1372008.pdf (last access: 19 July 2024), 2005.

Romeo, F., Porto, P., Mallamaci, C., and Muscolo, A.: The relationships between selected soil properties and caesium-137 identify organic carbon, nitrogen and water soluble phenols as indicators of soil erosion processes in different forest stands, J. Forestry Res., 32, 2589–2598, 2021.

Wischmeier, W. H. and Smith, D. D.: Predicting rainfall-erosion losses. A guide to conservation farming, United States Department of Agriculture Handbook, 537, 1–151, https://books.google.it/books?id=rRAUAAAAYAAJ&printsec=frontcover#v=onepage&q&f=false (last access: 19 July 2024), 1978.

Proc. IAHS, 387, 47–51, 2024
https://doi.org/10.5194/piahs-387-47-2024

The study of riparian areas in tourism: toward a conceptual framework of riparian tourism

Rajiv Dahal[1] **and Shamik Chakraborty**[2]

[1]Nepal Academy of Tourism and Hotel Management/Nepal Mountain Academy, Kathmandu, Nepal
[2]Graduate School of Advanced Sustainability Science, University of Toyama, Toyama, Japan

Correspondence: Rajiv Dahal (rajiv.dahal@gmail.com)

Published: 18 November 2024

Abstract. Riparian areas serve as interfaces between terrestrial and aquatic (including glaciers and glacial lakes) ecosystems, playing a crucial role in shaping landscapes, supporting flora and fauna diversity, and supporting human communities. Thus, riparian areas maintain ecological, cultural, and socio-economic resilience, enriching communities dependent on these ecosystems. Riparian areas are of great ecological value and have immense potential for tourism. However, the touristic value of riparian zones has largely remained unexplored and is confined mainly to the area of river-based recreational activities. This paper proposes "riparian tourism" as a holistic and sustainable form of tourism that encompasses both consumptive and non-consumptive forms of tourism. The exploration of the subject and the conceptualization of this potentially globally appealing form of tourism have the potential to offer entrepreneurial and touristic opportunities, especially for local communities, thereby ensuring not only ecological but also socio-economic benefits. The paper delves into creating a conceptual framework for riparian tourism, e.g. cryo-tourism. The research in this sense contributes greatly to increasing the discourse on sustainable tourism and emphasizes the urgency to incorporate tourism and conservation actions in riparian areas which are greatly impacted by the changing climate.

1 Introduction

"Riparian" is an anglicized form of the Latin word "riparius" (Naiman and Décamps, 1997), meaning land areas and the biotic communities that surround lakes, rivers, streams, and water bodies (see Naiman and Décamps, 1997; Singh et al., 2021; Verry et al., 2004; Tufekcioglu, 2000). Although many argue that riparian areas demonstrate traits of wetlands (Singh et al., 2021), Naiman and Décamps (1997) stated that, due to their lower frequency of saturation compared to traditional wetlands, they cannot be fully classified as wetlands. Rather, these are areas that encompass wetlands in their domain. As per Gregory et al. (1991), riparian zones serve as transitional areas between land and water ecosystems, with the authors proposing a conceptual model that integrates physical processes, terrestrial plant community succession, habitat formation, and the production of nutritional resources for aquatic ecosystems. In the same vein, riparian zones are identified as highly active components of the landscape (Swanson et al., 1988; Glass et al., 2015), which includes "important water resources, plant communities, fisheries, and wildlife" (Gregory et al., 1991).

Riparian areas are very important ecological resources because they control the flow of water and the quality of water (Anderson and Masters, 2023). These zones are important for water and landscape planning, aquatic ecosystem restoration, and building institutional and societal collaboration for these activities because of their extreme variety of species and environmental processes (Naiman and Décamps, 1997). The region being a hotspot of biodiversity serves as the sole reason for proposing these areas as tourist destinations. However, a conceptual framework that integrates tourism and recreation with the preservation of resources and with an emphasis on a region's socio-economic resilience is necessary for tourism in riparian areas.

2 Tourism in riparian zones

Riparian landscapes are "eco-symbols" because of their distinctive ecological and cultural characteristics, with a resemblance to features like hedgerows and paddy fields or to mountain features, embodying both ecological and symbolic significance for social groups and their surroundings (Berque et al., 1994, cited in Naiman et al., 2005). Furthermore, these landscapes, in the words of Jackson (1994), "come alive at certain periods of time" (Naiman et al., 2005, p. 239). In the same vein, Nassauer (1997) states that there is a dynamic interplay between natural environmental processes and cultural elements. Despite the great potential of resource-filled riparian areas, local communities living in and around these areas suffer from economic problems, and managers of these areas are burdened with balancing the economic benefits and social goals and ecological integrity. Therefore, when presenting riparian tourism, there are great opportunities for the development of small, medium, and micro enterprises (SMMEs), which helps greatly in job creation. This becomes essential for cultivating strong and beneficial relationships between communities and riparian areas.

3 Methodology

The objective of this paper is to propose a conceptual framework for riparian tourism, including cryo-tourism. In addition, the paper aims to analyse and advocate tourism opportunities in riparian areas, to adequately justify their resource opportunities, and to present strategies for showcasing the importance of integrating conservation efforts and tourism in these highly neglected areas. To meet the aims and objectives, the work employs a qualitative approach. We carry out a narrative literature review involving content analysis of the literature – peer-reviewed journal articles, conference proceedings and papers, and research work in the field of tourism and recreational and riparian areas – to explore the concept and facets of riparian tourism. The literature review helps to frame the meaning of riparian tourism and forms the basis for conceptualizing it.

4 Conceptualizing riparian tourism

Globally, riparian areas have proven to be full of resources for maintaining the ecological integrity of river landscapes. However, the touristic value of riparian areas has largely remained overlooked and detached from ecological, socio-economic, and conservation initiatives. Even though river and water resources have gained significant attention from the tourism industry, riparian areas' unique tourism potential has been insufficiently reported. This paper aims to conceptualize and bring attention to touristic opportunities in riparian areas, addressing the appeal, resources, and strategies that tourism could offer to enhance these areas and that have been overlooked in both academic and industry discussions.

Table 1. Aspects of riparian tourism.

Focus	Consumptive and non-consumptive forms of tourism
Scale	Multiple
Goal	Integration of recreation with conservation policies
Outcome	Conservation of riparian landscapes

Riparian tourism is holistic as well as broader and more resourceful than river tourism, lake tourism, or water-based tourism. In this sense, an effort has been made to elaborate on the meaning and concept of riparian tourism: *tourism is riparian-area-based when tourists travel to explore, understand, and appreciate riparian resources, thereby contributing economically and socially towards the conservation of such resources.*

To refine the meaning and concept of riparian tourism, four aspects and varieties are postulated (Table 1). Firstly, the focus of riparian tourism is on both consumptive and non-consumptive forms of tourism. Riparian areas have both consumptive and non-consumptive uses (Naiman et al., 2005). Non-consumptive uses include activities such as wildlife viewing, fishing, swimming, and natural aesthetics, whereas consumptive uses include consumptive activities that result in a decrease in the quality of environmental resources (Chen et al., 2019; Soman et al., 2007; Naiman et al., 2005). Riparian tourism is envisioned as having traits of both consumptive and non-consumptive forms of tourism.

From the scale point of view, there is a need for approaches, coordination, and scrutiny at the multiple levels of governance. The goal of riparian tourism is the integration of recreation with conservation policies. Finally, the outcome is the conservation and sustenance of riparian landscapes through tourism.

4.1 Riparian tourism framework

Riparian tourism centres around five themes and the interaction between them, i.e. (1) the environment (riparian resources), (2) community stewardship and solidarity, (3) institutional and public policy, (4) economic activity and rationality, and (5) recreational opportunities. The fundamental concept of riparian tourism is centred around the pristine and fragile riparian environment that forms resources for tourism. Besides their ecological value, riparian areas offer multiple recreational opportunities, such as hiking, cycling, golfing, horseback riding, bird watching, picnicking, camping, fishing, hunting, swimming, rafting, boating, and off-road vehicular travel (Zaimes, 2007). Thus, riparian tourism is one of the potential economic benefits of healthy riparian zones. The employment and diversified economic opportunities at the heart of the local-community approach make conceptualizing riparian tourism worthwhile. However, the success

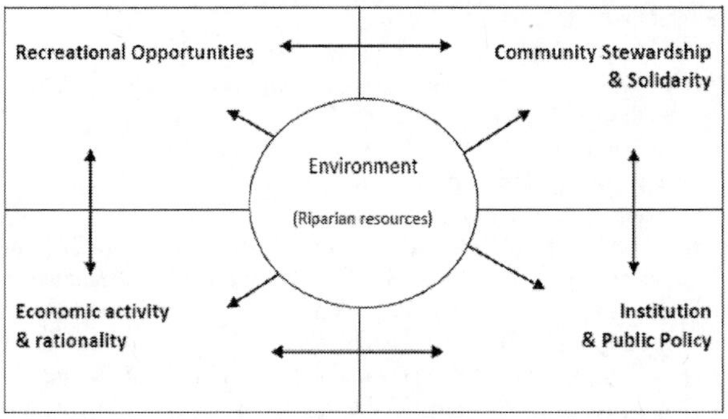

Figure 1. A schematic diagram of the riparian tourism framework.

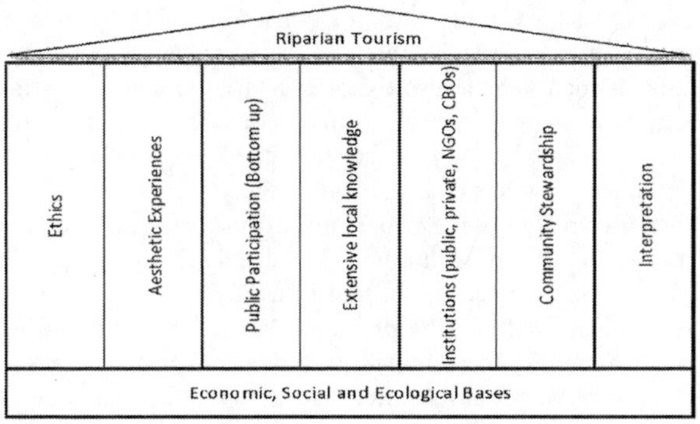

Figure 2. The seven pillars of sustainable riparian tourism.

of such riparian tourism hinges pretty much on community stewardship and solidarity becoming integral to the framework. Both community and riparian resources have benefitted from each other since time immemorial.

Communities have been protecting their resources, either because these resources support their livelihood or because these places are symbols of spiritual and cultural bonds. Riparian resources provide clean air, water, and arable land for economic vitality to these traditional communities. Communities have long been recognized as stewards of riparian resources. Any inclination towards designing and developing riparian tourism should be centred around "community ownership and stewardship", which is advocated as a cornerstone of institutional mandates and public-policy discourse. A review of the literature from Nepal, Japan, and around the globe shows that there is limited discussion of the facets of riparian areas and tourism in governmental and policy-making directives. A robust institutional mandate (either added to the current supervisory organizational set-up for water, forests, and glaciers or, ideally, separated from current bureaucratic forms) and a separate policy on riparian areas and tourism are of paramount importance. The arrows in Fig. 1 show the multi-headed dependence and influence of each factor on the others when working towards the sustenance of resources and tourism alike.

The associated impact of tourism and recreational use is immense in the fragile and exotic ecosystem, thus drawing the attention of policy-makers and planners to the sustainability of these important ecosystems. The sustainability of these resources plays a significant role in the sustainability of riparian tourism itself. The section below presents a brief discussion of the sustainability of riparian tourism.

4.2 Sustainability – riparian tourism

Riparian areas are considered useless, drained, and destroyed or are used as disposal areas (Naiman et al., 2005). Humans have long intervened in the riparian system, mainly for agriculture (such as vineyards, unsustainable occupation, or unsustainable development such as concretization and compaction) and recreation (Arif et al., 2023; Boucher and Hudson, 2023. Zingraff-Hamed et al., 2022; Zaimes and Iakovoglou, 2020; Zaimes, 2007). The cost of preserving the riparian system is very high compared to the economic value generated by humans through agriculture, recreation, and other activities (Lind et al., 2019; Stutter et al., 2019). Regarding the use of areas for recreation, Winter (1993) pointed out that, in the West, "recreationists" are lured to riparian areas because of water resources and lush vegetation. Contrary to popular belief, Nassauer (1997) argues that riparian landscapes, when they offer aesthetic experiences, stand a greater chance of conservation within the human-dominated world provided they evoke sustained interest in people (Naiman et al., 2005).

From the work of Nassauer (1992, 1997), it is evident that the sustainability of riparian areas depends on governance, management, and control, with a strong reliance on humans. Such dependence on humans requires active and meaningful attention to be paid to these landscape features (Nassauer, 1992). In this sense, the foundation of the sustainability of riparian tourism remains the economic, socio-cultural, and ecological integrity of riparian resources. Seven pillars for holistic development and promotion of riparian tourism are proposed, i.e. ethics, aesthetic experiences, public participation (bottom-up), extensive local knowledge use, institutions (public, private, non-governmental, and community-based organizations), community stewardship, and interpretation (Fig. 2).

5 Conclusions

The conservation of riparian areas is an invaluable contribution to wildlife, communities, and the environment and includes stabilizing and restoring riverbanks, supporting biodiversity, and increasing economic resilience. Riparian tourism integrated with local socio-cultural and environmental resources can help locals with much-needed cash flow, supporting both survival and resource conservation. Despite their economic value, recreational activities like camping, hiking, and off-road vehicle use threaten riparian systems, yet research on the prospects and impact of riparian tourism remains unexplored. Nepal and other developing nations have yet to fully assess and capitalize on the prospect of this new tourism type. Thus, this paper aims to introduce and advocate this novel concept with a focus on sustainability in riparian tourism to stakeholders, primarily policy-makers, planners, researchers, and industry practitioners. The model proposed centres around the community's ownership and stewardship, furthering robust institutional set-ups and active public-policy discourse. Placing ecological ethics and aesthetic experiences at the centre of riparian tourism, with a strong emphasis on the interpretation of these resources, frames the direction for further exploration and description. The value tourism brings to these fragile yet neglected resources is important not only for conservation, but also for supporting the resilience of the local economy and, ultimately, that of the community, who are the actual stewards of these resources.

Data availability. No data sets were used in this article.

Author contributions. RD and SC did the conceptualization. RD wrote the manuscript with considerable input from SC. Both the authors reviewed and edited the paper.

Competing interests. The contact author has declared that neither of the authors has any competing interests.

Disclaimer. Publisher's note: Copernicus Publications remains neutral with regard to jurisdictional claims made in the text, published maps, institutional affiliations, or any other geographical representation in this paper. While Copernicus Publications makes every effort to include appropriate place names, the final responsibility lies with the authors.

Special issue statement. This article is part of the special issue "Mountain Hydrology and Cryosphere". It is a result of the International Conference on Mountain Hydrology and Cryosphere, Kathmandu and Dhulikhel, Nepal, 9–10 November 2023.

Acknowledgements. We thank the reviewers for their comments and suggestions, which made this paper a better one.

References

Anderson, S. and Masters, R.: Riparian Forest Buffers, Oklahoma State University, Oklahoma, https://extension.okstate.edu/fact-sheets/print-publications/nrem/water-quality-series-riparian-forest-buffers-nrem-5034.pdf (last access: 10 January 2024), 2023.

Arif, M., Jie, Z., Behzad, H. M., and Changxiao, L.: Assessing the impacts of ecotourism activities on riparian health indicators along the Three Gorges Reservoir in China, Ecol. Indic., 155, 111065, https://doi.org/10.1016/j.ecolind.2023.111065, 2023.

Berque, A., Conan, M., Donadieu, P., Lassus B., and Roger A.: Cinq propositions pour une théorie du paysage, Champ Vallon, ISBN: 978-2876731912, 1994.

Boucher, Z. and Hudson, P. F.: Troubled waters: Riparian ecosystem services and community opposition to the largest dam removal project in Europe, Vezins Dam, France, Geoforum, 147, 103906, https://doi.org/10.1016/j.geoforum.2023.103906, 2023.

Chen, W. Y., Li, X., and Hua, J.: Environmental amenities of urban rivers and residential property values: a global meta-analysis, Sci. Total Environ., 693, 133628, https://doi.org/10.1016/j.scitotenv.2019.133628, 2019.

Glass, J. R. and Floyd, C. H.: Effects of proximity to riparian zones on avian species richness and abundance in montane aspen woodlands, J. Field Ornithol., 86, 252–265, https://doi.org/10.1111/jofo.12105, 2015.

Gregory, S. V., Swanson, F. J., McKee, W. A., and Cummins, K. W.: An ecosystem perspective of riparian zones, BioScience, 41, 540–551, https://doi.org/10.2307/1311607, 1991.

Huppert, D. and Kantor, S.: Economic perspectives, in: River Ecology and Management – lessons from the Pacific Coastal Ecoregion, edited by: Naiman, R. J. and Bilby, R. E., Springer-Verlag, New York, 572–595, https://doi.org/10.1007/978-1-4612-1652-0_23, 1998.

Jackson, J. B.: A sense of place, a sense of time, Yale University Press, New Haven, CT, 212, 0300060025, ISBN: 978-0300063974, 1994.

Lind, L., Hasselquist, E. M., and Laudon, H.: Towards ecologically functional riparian zones: A meta-analysis to develop guidelines for protecting ecosystem functions and biodiversity in agricultural landscapes, J. Environ. Manage., 249, 109391, https://doi.org/10.1016/j.jenvman.2019.109391, 2019.

Naiman, R. J. and Decamps, H.: The ecology of interfaces: riparian zones. Annu. Rev. Ecol. Syst., 28, 621–658, https://doi.org/10.1146/annurev.ecolsys.28.1.621, 1997.

Naiman, R. J., Decamps, H., and McClain, M. E.: Riparia: Ecology, Conservation, and Management of streamside communities, BioScience, 56, 353–354, https://doi.org/10.1641/0006-3568(2006)56[353:FL]2.0.CO;2, 2005.

Nassauer, J. I.: The appearance of ecological systems as a matter of policy, Landsc. Ecol., 6, 239–250, https://doi.org/10.1007/bf00129702, 1992.

Nassauer, J. I. (Ed.): Placing Nature: Culture and Landscape Ecology, Island Press, Washington, DC, 215, 1559635592, ISBN 978-1559635592, 1997.

Rasmussen, G. A. and Padgett, W.: Recreational effects on riparian areas, Nat. Resour. Env. Iss., 1, 7, 45–48, https://digitalcommons.usu.edu/nrei/vol1/iss1/7 (last access: 17 August 2024), 1994.

Singh, R., Tiwari, A. K., and Singh, G. S.: Managing riparian zones for river health improvement: an integrated approach, Landscape Ecol. Eng., 17, 195–223, https://doi.org/10.1007/s11355-020-00436-5, 2021.

Stutter, M., Kronvang, B., Ó hUallacháin, D., and Rozemeijer, J.: Current insights into the effectiveness of riparian management, attainment of multiple benefits, and potential technical enhancements, J. Environ. Qual., 48, 236–247, https://doi.org/10.2134/jeq2019.01.0020, 2019.

Swanson, F. J., Kratz, T. K., Caine, N., and Woodmansee, R. G.: Landform effects on ecosystem patterns and processes, BioScience, 38, 92–98, https://doi.org/10.2307/1310614, 1988.

Tufekcioglu, A.: Biomass, carbon, nitrogen and soil respiration dynamics within riparian buffers and adjacent crop fields. Iowa State University, Ames, Iowa, 104 pp., https://doi.org/10.31274/rtd-180813-13637, 2000.

Verry, E. S., Dolloff, C. A., and Manning, M. E.: Riparian ecotone: a functional definition and delineation for resource assessment, Water Air. Soil Poll.: Focus, 4, 67–94, https://doi.org/10.1023/B:WAFO.0000012825.77300.08, 2004.

Winter, P. L.: Positives and negatives of recreation in riparian areas, Riparian Management: Common Threads and Shared Interests: A Western Regional Conference on River Management Strategies: February 4–6, 1993, Albuquerque, New Mexico, US Department of Agriculture, Forest Service, Rocky Mountain Forest and Range Experiment Station, 226, 155-158, https://www.fs.usda.gov/rm/boise/AWAE/labs/awae_flagstaff/Hot_Topics/ripthreatbib/winter_posnegrecripareas.pdf (last access: 20 January 2024), 1993.

Zaimes, G.: Human alterations to riparian areas, in: Understanding Arizona's Riparian Areas, edited by Zaimes, G., College of Agriculture and Life Sciences, The University of Arizona, Tucson, AZ, az1432, 83–109, https://rangelandsgateway.org/sites/default/files/2021-01/az1432.pdf, (last access: 12 January 2024), 2007.

Zaimes, G. N. and Iakovoglou, V.: Assessing riparian areas of Greece - an Overview,Sustainability, 13, 309, https://doi.org/10.3390/su13010309, 2020.

Zingraff-Hamed, A., George, F. N., Lupp, G., and Pauleit, S.: Effects of recreational use on restored urban floodplain vegetation in urban areas, Urban For. Urban Green., 67, 127444, https://doi.org/10.1016/j.ufug.2021.127444, 2022.

Proc. IAHS, 387, 53–58, 2024
https://doi.org/10.5194/piahs-387-53-2024

Open Access PIAHS

Hydropower potential of the Marsyangdi River and Bheri River basins of Nepal and their sensitivity to climate variables

Rakesh Kayastha[1], **Rijan Bhakta Kayastha**[1], **Kundan Lal Shrestha**[2], **and Smriti Gurung**[3]

[1]Himalayan Cryosphere, Climate and Disaster Research Centre (HiCCDRC), Department of Environmental Science and Engineering, School of Science, Kathmandu University, Dhulikhel, Nepal

[2]Department of Chemical Science and Engineering, School of Engineering, Kathmandu University, Dhulikhel, Nepal

[3]Department of Environmental Science and Engineering, School of Science, Kathmandu University, Dhulikhel, Nepal

Correspondence: Rakesh Kayastha (rakesh.kayastha@ku.edu.np)

Published: 18 November 2024

Abstract. Understanding the hydrology of the Himalayan region and its response to current and future climate scenarios is crucial for identifying the region's future water availability for infrastructural development. For this, the hydropower potential and the impact of future change on the hydrology of the Bheri River basin (BRb) and the Marsyangdi River basin (MRb) were analysed. The Glacio-hydrological Degree-Day Model version 2.0 (GDM V2.0), developed in the PCRaster dynamic modelling framework, was used to simulate the river runoff. Geospatial tools and different criteria were used to assess topographic features, identify suitable places for run-of-river (ROR) hydropower development, and estimate power potential. Eight scenarios with various combinations of temperature and precipitation changes were used to assess the hydropower potential. Increases in temperature by 0.5 and 1 °C (assumed for the near-term and mid-term future) and changes in precipitation of ± 10 % and ±20 %, respectively, are used to conduct a sensitivity analysis for the hydropower potential. A total of 116 and 83 suitable sites were identified, and 4242 and 2823 MW power potentials were estimated in the BRb and the MRb, respectively. All the sensitivity scenarios show an increase in hydropower production, except for one with a drier scenario and less precipitation. The integration of a geographic information system (GIS) and a hydrological model helps us to understand the hydrological response to climate variables and its impact on hydropower in the Himalayan region.

1 Introduction

Understanding the hydrological characteristics of glacierized catchments is critical for assessing the hydrological dynamics of basins. For communities downstream, these areas provide the majority of the water resources needed for hydropower, irrigation, water supply, and livelihood. Local or global climate variability can have a direct impact on the most efficient use of water resources, particularly hydropower. The variations in climate variables affect the patterns of river runoff, which in turn affects the amount of water available for the production of hydroelectric energy (Immerzeel et al., 2013; Lutz et al., 2014). However, climate change may alter hydrological variability, affecting the water resource systems that depend on the flow regime. Because of the complex connection between variations in the climate, mountainous topography, and hydrological processes and because of the insufficient data, the potential effects of changes in the hydroclimate on hydropower generation are not generally understood, particularly in a Nepalese context. Previously, hydropower infrastructure, design, and operation were estimated using historical streamflow records as a source of information (Shrestha et al., 2014). Shrestha (1966) estimated a total hydropower potential of 83 500 MW in

Nepal, with an economically feasible capacity of 45 610 MW from 114 sites. Additionally, a recent study suggested that Nepal has a run-of-river (ROR) hydropower potential of 53 836 MW with a 40 % dependable flow (Jha, 2010; Pradhan, 2009; Shrestha, 1966; WECS, 2011).

Numerous modelling techniques have been developed to enhance comprehension of hydrological dynamics with different levels of complexity. The hydrological model integrated with the geographic information system (GIS) offers a suitable application for hydropower potential evaluation. The hydropower capacity can be estimated more effectively with this integration.

The objective of this study is to assess suitable sites for hydropower development and its power potential in the Bheri River and Marsyangdi River basins using an integrated modelling approach.

2 Study area

This study focuses on the two glacierized sub-basins of the Karnali River and Gandaki River systems, the Bheri River basin (BRb) and the Marsyangdi River basin (MRb), located in western Nepal (Fig. 1). The BRb elevation ranges from 227 to 7598 m above sea level (a.s.l.) with a total area of 13 688 km^2, and it is one of the major tributaries of the Karnali River, the third largest river system in Nepal. About 3.93 % (538.71 km^2) of the total area of the BRb is glacierized; within that, debris-covered glaciers cover 28.43 km^2 (0.02 %). While the MRb is the major tributary of the Gandaki River system, elevation ranges from 258 to 7593 m a.s.l., with a total area of 4786 km^2, where 16.75 % (802 km^2) of the basin is covered by debris-free glaciers and 0.59 % (28 km^2) is covered by debris-covered glaciers. The higher regions of the BRb and MRb lie on the leeward side of mountain ranges (the Dhaulagiri and Annapurna massifs) in the trans-Himalayan region.

The steep slopes and high potential water energy due to the mountainous landscapes in both basins offer opportunities for developing hydropower projects. As of 2023, the Department of Electricity Development (DOED), a government agency in Nepal, has granted licenses for 56 hydropower projects in the MRb. These projects have a combined capacity to generate 2007 MW of electricity. Of these, four hydropower projects in the MRb have a capacity of more than 100 MW of electricity production. In the BRb, 33 hydropower projects have a combined electricity generation capacity of 3934 MW. Notably, 10 of these projects have a capacity exceeding 100 MW. The basin's main rivers, the Marsyangdi and Bheri, have significant hydropower projects because of their consistently high runoff throughout the year.

3 Input data

3.1 Meteorological and hydrological data

Input data such as daily air temperature, precipitation, and discharge data were obtained from the Department of Hydrology and Meteorology (DHM), Government of Nepal, for the hydrological analysis. The meteorological data with maximum and minimum air temperatures as well as the amount of precipitation were recorded at eight and six stations, respectively, at varying elevations within the basin boundary (Table 1). Daily discharge data from the hydrological stations at Samaijighat at an elevation of 500 m a.s.l. and Bimalnagar at an elevation of 354 m a.s.l. were used for the BRb and the MRb, respectively. The available observed time series meteorological data were used as a forcing dataset for calibrating and validating the hydrological model simulations.

3.2 Spatial data

For the hydrological modelling, Environmental Systems Research Institute (ESRI) Land Use Land Cover (LULC) derived from 10 m resolution Sentinel-2 imagery is obtained from ESRI (https://livingatlas.arcgis.com/landcoverexplorer, last access: November 2022) (Karra et al., 2021) for the land cover information. Ten different class definitions from the ESRI LULC dataset are modified and reclassified with similar topographic features and surface runoff behaviour to create eight land classes for a similar range of rainfall-runoff coefficients. Similarly to this, the Shuttle Radar Topography Mission (SRTM) 1 arcsec global digital elevation model (DEM) with 30 m spatial resolution is obtained from the USGS (https://earthexplorer.usgs.gov, last access: January 2023) and serves as the grid and elevation data for the hydrological model and hydropower potential. Randolph Glacier Inventory Version 6.0 (RGI v6.0) (Arendt et al., 2017) glacier outlines are used for the glacier information and for reclassifying land classes.

3.3 Climate sensitivity

This study considered various scenarios to simulate the potential effects of climate variation on the basins' hydrology. These scenarios were based on changing the temperature and precipitation inputs to the hydrological model following various assumptions about greenhouse gas emissions and their effects. In the Ganga basin, Mishra et al. (2020) projected that temperatures will change by approximately 0.55 ± 0.27 °C in the near future (NF) and by 1.37 ± 0.42 °C in the mid-future (MF), while precipitation is expected to change by about 6.95 ± 7.78 % in the NF and by 15.05 ± 11.3 % in the MF under the SSP245 scenarios.

To represent the NF and MF scenarios, the temperature is increased by 0.5 and 1 °C, respectively. For drier and wetter conditions, precipitation input values are changed by adding

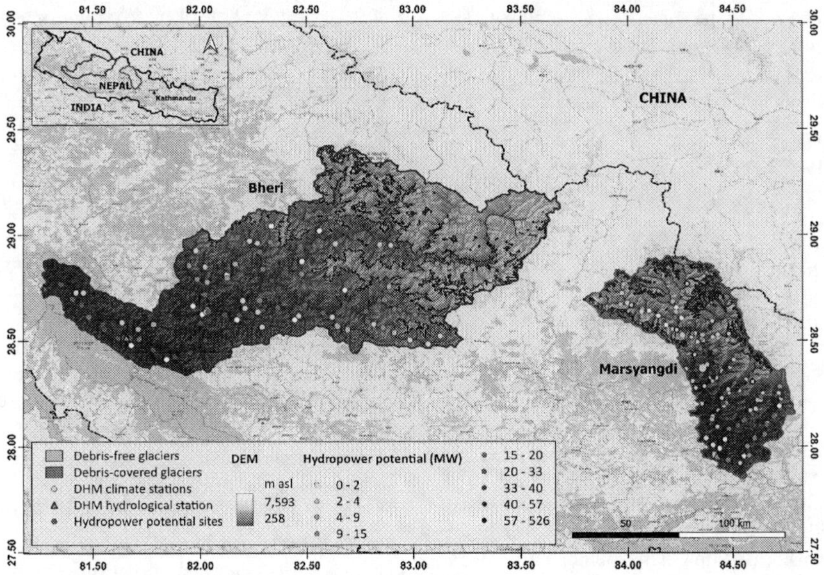

Figure 1. Locations of the Marsyangdi River and Bheri River basins in Nepal (Nepal is in the inset). The black line indicates the Nepalese political boundary, the points indicate the hydropower project's potential sites, and the colour ramp indicates its power potential. The green triangle and yellow points represent the hydrological and climatological stations of the Department of Hydrology and Meteorology (DHM) in the MRb and BRb. Background: © OpenStreetMap contributors 2024. Distributed under the Open Data Commons Open Database License (ODbL) v1.0.

Table 1. List of the climatological and hydrological stations in the study area.

Basin	Station (m a.s.l.), type	Hydrological station	Year
Bheri River	Dunai (2098), precipitation*; Mainagau (1913), precipitation; Rukumkot (1568), precipitation; Musikot (1412), climatology; Jagarkot (1240), climatology; Chaurjhar (863), climatology*; Surkhet (720), synoptic	Samaijighat	2004–2017
Marsyangdi River	Phugau (4100), precipitation; Manangbhot (3420), climatology; Chame (2680), climatology*; Taal (1670), precipitation; Faleni (1278), precipitation; Khudibazar (823), climatology*	Bimalnagar	2004–2010

* Reference station for the respective basins.

and subtracting 10 % and 20 %, respectively, from the baseline values. These temperature and precipitation changes produced eight different climate scenarios (Table 2), which are used as forcing data for the hydrological model and to simulate future hydropower potential.

4 Methodology

4.1 Glacio-hydrological model

The Glacio-hydrological Degree-Day Model version 2.0 (GDM V.2) is applied in this study to estimate the discharge of the rivers. It is a distributed glacio-hydrological model developed in the PCRaster modelling framework that can estimate the contributions from snowmelt, ice melt, rain, and baseflow to the river discharge from the glacierized basins (Kayastha et al., 2020; Kayastha and Kayastha, 2019; Khadka et al., 2020; Yang et al., 2022). GDM incorporates a variety of hydrological processes, including groundwater hydrological processes, cryospheric processes, rainfall runoff, and evapotranspiration. The melt module in GDM uses the degree-day approach, a simplified version of the intricate procedure (Braithwaite and Olesen, 1989) for estimating glacier ice melt and snowmelt with the least amount of data needed (Kayastha et al., 2005; Pradhananga et al.,

Table 2. Climate scenarios with the combinations of climate variables to project the future hydropower potential.

Combination of climate variables*	Scenario
+1 °C + 10 %	S1
+1 °C + 20 %	S2
+1 °C − 10 %	S3
+1 °C − 20 %	S4
+0.5 °C + 10 %	S5
+0.5 °C + 20 %	S6
+0.5 °C − 10 %	S7
+0.5 °C − 20 %	S8

* Temperature (°C) and precipitation (%).

2014). This approach calculated the melts from debris-free, debris-covered and snow above the glaciers as well as snow.

In this study, the 30 m DEM and 10 m LULC were resampled to a 180 m grid, and then GDM was forced for hydrological simulation in both basins. The meteorological data are pre-processed to force the model for the simulation. The daily precipitation is distributed on the resampled grid using the inversed distance weighting (IDW) approach to interpolate precipitation from the meteorological stations to the grid points. The elevation-dependent temperature distribution in the basin is derived by calculating the monthly temperature lapse rate between the reference stations and the highest-elevation station (∗ in Table 1). An environmental lapse rate ($-6\,°\mathrm{C\,km^{-1}}$) is used to distribute temperature above the higher elevation to the rest of the grids.

The threshold temperature determines whether the precipitation is in the form of snow or rain in the grids. The melts from the debris-free and debris-covered glaciers and snow are calculated with the degree-day factor from each grid. Different sets of degree-day factors are applied for each month, covering the range of estimated degree-day factors determined by various glaciers in the Nepalese Himalayas. For the snowmelt and ice-melt processes, two distinct sets of degree-day factors are used; lower factors are applied at elevations below 5000 m, and high factors are applied at elevations above 5000 m. The model accounts for seasonal variations in melting rates, with higher rates in non-monsoon months (reflected by higher degree-day factors) and lower rates during the monsoon.

The daily time series of observed and simulated discharges are compared to determine the model's performance efficiency using the Nash–Sutcliffe efficiency (NSE) and volume difference (VD) indices (Kayastha et al., 2020). The NSE index plays a crucial role in assessing model accuracy in this study. A value exceeding 0.7, coupled with a volume difference within ±10 %, indicates high accuracy and certainty.

4.2 Identification of hydropower sites and potential estimation

The geospatial tools in Quantum Geographic Information System (QGIS) are used to identify potential sites for hydropower projects. The pre-processing steps for the hydrological analysis, such as filling voids, flow directions, and accumulation, were performed for the stream networks and watersheds. The third order of the stream hierarchy with the Strahler methods was evaluated as a potential location for hydropower development, ensuring a consistent water flow. A minimum distance of 7 km between the two consecutive hydropower plants is considered to provide adequate space for civil works and environmental considerations, such as reservoirs, powerhouses, and environmental flows. For the steepness, high gradient, and altitude, an elevation below 4000 m is considered feasible for hydropower project development. The gross head of the stream segments is extracted using the DEM at the initial and final points. In this study, 70 % of the gross head is considered the net or effective head, which will be used for potential estimation. The head greater than 20 m is only considered in this study. The run-of-river hydropower potential is calculated with the following relationship (Eq. 1):

$$P\,(\mathrm{MW}) = \frac{\rho \cdot g \cdot Q \cdot h}{10^6}, \qquad (1)$$

where ρ is the density of water ($1000\,\mathrm{kg\,m^{-3}}$), g is the gravitational acceleration ($9.81\,\mathrm{m\,s^{-2}}$), Q is the discharge ($\mathrm{m^3\,s^{-1}}$), and h is the net or effective head (m).

The daily runoff at potential sites from the hydrological model is further examined to determine the dependable flow. For a potential hydropower site, the flow exceedance or design discharge is considered at the 40th percentile of dependable flow (Q_{40}), which is calculated using the flow duration curve (FDC).

5 Results and discussion

5.1 Calibration and validation

The model's performance is first evaluated in two river basins by comparing the simulated discharge to the observed discharge in the specific period: from 2004 to 2017 in the BRb and from 2004 to 2010 in the MRb, using observed discharge data from the hydrological station in the basin shown in Fig. 2a and b, respectively. In the BRb, the model's performance is evaluated using 13-year runoff data from the hydrological station at Samaijighat. The model is validated from 2011 to 2017 and calibrated for 6 years from 2004 to 2010. Similarly, the MRb made use of 7 years of runoff data from the Bimalnagar hydrological station. Three years (2008–2010) of observed runoff are used for validation and 4 years (2004–2007) for calibration in the MRb. The NSE and VD metrics of the model's performance indices are used to assess the model's performance. In the BRb and the MRb,

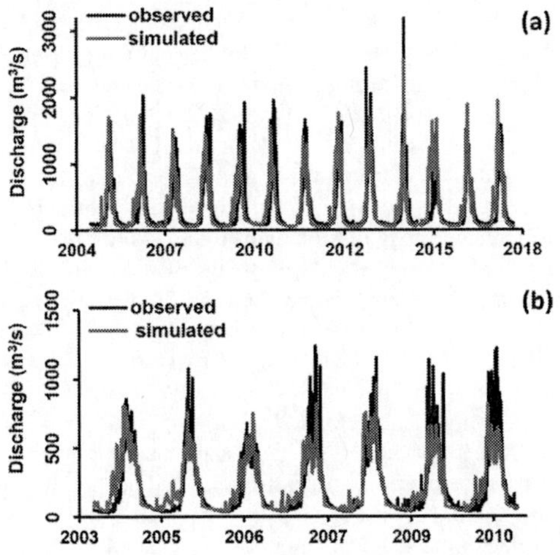

Figure 2. The comparison of observed and simulated runoff hydrographs during calibration and validation in **(a)** the BRb and **(b)** the MRb.

the NSEs for the calibration period are 0.82 and 0.81, and the VDs are 4.3 and −5.8, respectively. The NSE for the validation period for the BRb is 0.71, with a VD of −5.3. Similarly, these are 0.7 and 10.6 in the MRb for the validation period. The runoff calibration and validation performance in both of the basins are considered good.

5.2 Suitable sites and hydropower potential

The geospatial analysis identified possible locations for hydropower development in the BRb and the MRb. Overall, 116 potential sites in the BRb and 83 sites in the MRb were identified. Potential sites in the BRb have a catchment area ranging from 17.95 to 865.2 km^2, while in the MRb this varies from 14.93 to 316.57 km^2. Similarly, in the BRb, the net or effective head varies from 21 to 827 m, whereas in the MRb it varies from 23 to 1034 m.

The gross hydropower potential at each appropriate location was determined using the runoff from the hydrological model simulation. The Q_{40} from the FDC, the net or effective head, was used to determine the hydropower potential at each appropriate location. According to current climate estimates, the BRb has a gross hydropower production capacity of 4142 MW at 40 % flow exceedance compared to 2823 MW for the MRb. Jha (2010) estimated a 4140 MW capacity in the BRb and a 3251 MW capacity in the MRb using the hydropower model. Similarly, Bajracharya (2015) estimated total potentials of 7755 MW in the BRb and 8257 MW in the MRb using the mean annual flow and 30 % of the dependable flow. WECS (2019) reported higher potentials of 4614 MW in the MRb and 7993 MW in the BRb, which were estimated between the sub-basin level and the adjacent river confluence. This study adopts the concept of cascading, or

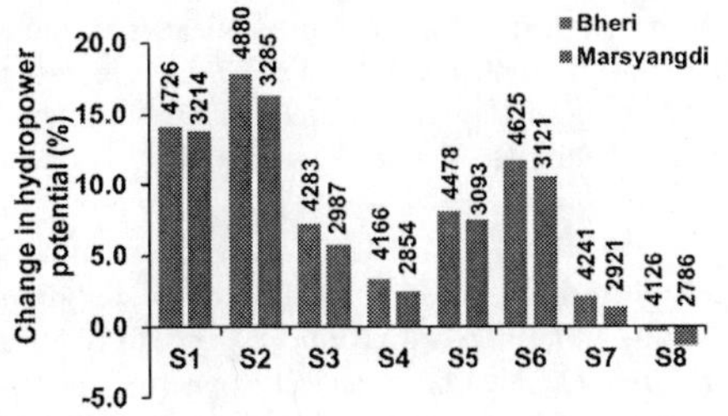

Figure 3. This figure shows the relative change in hydropower potential (%) from the present climate condition and the total power potential (MW) above the bars in various climate sensitivities.

sequential development, of hydropower projects along rivers and uses the same water flow multiple times to produce electricity.

Although there were only 5 out of 83 potential sites in the MRb, out of 116, 15 potential sites were found to be able to produce more than 100 MW in the BRb. Small hydropower, defined as less than 10 MW, is categorized as 35 in the MRb and as 55 in the BRb. A sensitivity analysis of climate variables revealed that the potential for hydropower increases in all the settings, except for the extremely dry scenario (S8) (Fig. 3).

6 Conclusion

An integrated approach is used to evaluate hydropower potential and locate potential run-of-river hydropower locations in Nepal's two river basins. As the methodological framework, the GIS and GDM are integrated. The observed daily runoff is used to calibrate and validate the hydrological model at daily time steps, and the NSE and VD indices are used to assess the model's performance. The geospatial criteria and tools in the GIS are used to identify possible locations for hydropower development in the future: 116 and 83 suitable sites are identified in the BRb and MRb, with total hydropower production capacities of 4142 and 2823 MW, respectively. The sensitivity of the hydropower potential is assessed by varying the temperature and precipitation relative to the baseline data. Hydropower production will increase in all the sensitivity scenarios, except for one in which precipitation decreases by 20 %. Hydropower developers, planners of water resources, and legislators can maximize the use of water resources in the Himalayan basin with the help of these findings. Moreover, the study offers insights that can guide future academic investigations into key factors influencing hydropower potential.

Code and data availability. The code for the glacio-hydrological model as well as the data supporting the findings of this study are available from the corresponding author on request.

Author contributions. RK: conceptualization, methodology, analysis, writing – original draft. RBK: GDM concept, supervision, review and editing. KLS: GDM development, review and editing. SG: review and editing.

Competing interests. At least one of the (co-)authors is a guest member of the editorial board of *Proceedings of IAHS* for the special issue "Mountain Hydrology and Cryosphere". The peer-review process was guided by an independent editor, and the authors also have no other competing interests to declare.

Disclaimer. Publisher's note: Copernicus Publications remains neutral with regard to jurisdictional claims made in the text, published maps, institutional affiliations, or any other geographical representation in this paper. While Copernicus Publications makes every effort to include appropriate place names, the final responsibility lies with the authors.

Special issue statement. This article is part of the special issue "Mountain Hydrology and Cryosphere". It is a result of the International Conference on Mountain Hydrology and Cryosphere, Kathmandu and Dhulikhel, Nepal, 9–10 November 2023.

Financial support. This research has been supported by the University Grant Commission – Nepal (UGC COVID-19 Response, Technical Innovation, Scientific Investigation and Research Grant 2077/78).

References

Arendt, A., Bliss, A., Bolch, T., et al.: Randolph glacier inventory – A dataset of global glacier outlines: Version 6.0: Technical report, global land ice measurements from space, RGI consortium, https://doi.org/10.7265/4m1f-gd79, 2017.

Bajracharya, I.: Assessment of run-of-river hydropower potential and power supply planning in Nepal using hydro resources, Inst. für Energietechnik und Thermodyn, 7–12, https://doi.org/10.34726/hss.2015.27916, 2015.

Braithwaite, R. J. and Olesen, O. B.: Calculation of Glacier Ablation from Air Temperature, West Greenland, in: Glacier Fluctuations and Climatic Change, Springer, Dordrecht, the Netherlands, 219–233, 1989.

Immerzeel, W. W., Pellicciotti, F., and Bierkens, M. F. P.: Rising river flows throughout the twenty-first century in two Himalayan glacierized watersheds, Nat. Geosci., 6, 742–745, https://doi.org/10.1038/ngeo1896, 2013.

Jha, R.: Total Run-of-River type Hydropower Potential of Nepal, Hydro Nepal J. Water, Energy Environ., 7, 8–13, https://doi.org/10.3126/hn.v7i0.4226, 2010.

Karra, K., Kontgis, C., Statman-Weil, Z., Mazzariello, J. C., Mathis, M., and Brumby, S. P.: Global land use/land cover with Sentinel 2 and deep learning, in: 2021 IEEE international geoscience and remote sensing symposium IGARSS, 4704–4707, https://doi.org/10.1109/IGARSS47720.2021.9553499, 2021.

Kayastha, R. B. and Kayastha, R.: Glacio-hydrological degree-day model (gdm) useful for the himalayan river basins, Himal. Weather Clim. their Impact Environ., Springer, Cham, 379–398, https://doi.org/10.1007/978-3-030-29684-1_19, 2019.

Kayastha, R. B., Ageta, Y., and Fujita, K.: Use of positive degree day methods for calculating snow and ice melting and discharge in glacierized basins in the Langtang Valley, Central Nepal, Climate and Hydrology in Mountain Areas, 7–14, 2005.

Kayastha, R. B., Steiner, N., Kayastha, R., Mishra, S. K., and McDonald, K.: Comparative Study of Hydrology and Icemelt in Three Nepal River Basins Using the Glacio-Hydrological Degree-Day Model (GDM) and Observations From the Advanced Scatterometer (ASCAT), Front. Earth Sci., 7, 354, https://doi.org/10.3389/feart.2019.00354, 2020.

Khadka, M., Kayastha, R. B., and Kayastha, R.: Future projection of cryospheric and hydrologic regimes in Koshi River basin, Central Himalaya, using coupled glacier dynamics and glacio-hydrological models, J. Glaciol., 66, 831–845, https://doi.org/10.1017/jog.2020.51, 2020.

Lutz, A. F., Immerzeel, W. W., Shrestha, A. B., and Bierkens, M. F. P.: Consistent increase in High Asia's runoff due to increasing glacier melt and precipitation, Nat. Clim. Chang., 4, 587–592, https://doi.org/10.1038/nclimate2237, 2014.

Mishra, V., Bhatia, U., and Tiwari, A. D.: Bias-corrected climate projections for South Asia from Coupled Model Intercomparison Project-6, Sci. Data, 7, 1–13, https://doi.org/10.1038/s41597-020-00681-1, 2020.

Pradhan, P. M. S.: Hydropower development, in: The Nepal–India Water Relationship: Challenges, Springer, 125–151, https://doi.org/10.1007/978-1-4020-8403-4_5, 2009.

Pradhananga, N. S., Kayastha, R. B., Bhattarai, B. C., Adhikari, T. R., Pradhan, S. C., Devkota, L. P., Shrestha, A. B., and Mool, P. K.: Estimation of discharge from Langtang River basin, Rasuwa, Nepal, using a glacio-hydrological model, Ann. Glaciol., 55, 223–230, https://doi.org/10.3189/2014AoG66A123, 2014.

Shrestha, H. M.: Cadastre of Potential Hydropower Resources in Nepal, Moscow Power Institute, 1966.

Shrestha, S., Khatiwada, M., Babel, M. S., and Parajuli, K.: Impact of Climate Change on River Flow and Hydropower Production in Kulekhani Hydropower Project of Nepal, Environ. Process., 1, 231–250, https://doi.org/10.1007/s40710-014-0020-z, 2014.

WECS: Water Resources of Nepal in the context of Climate Change, Technical report, Water and Energy Commission Secretariat (WECS), 2011.

WECS: Assessment of Hydropower Potential of Nepal, Technical report, Water and Energy Commission Secretariat (WECS), 2019.

Yang, M., Li, Z., Anjum, M. N., Kayastha, R., Kayastha, R. B., Rai, M., Zhang, X., and Xu, C.: Projection of Streamflow Changes Under CMIP6 Scenarios in the Urumqi River Head Watershed, Tianshan Mountain, China, Front. Earth Sci., 10, 1–14, https://doi.org/10.3389/feart.2022.857854, 2022.

Proc. IAHS, 387, 59–63, 2024
https://doi.org/10.5194/piahs-387-59-2024

Open Access
PIAHS

Glacial lake outburst flood (GLOF) modeling of Tsho Rolpa glacial lake, Nepal

Rijan Bhakta Kayastha[1] **and Sunwi Maskey**[2]

[1]Himalayan Cryosphere, Climate and Disaster Research Center, Department of Environmental Science and Engineering, School of Science, Kathmandu University, Dhulikhel, Nepal
[2]International Centre for Integrated Mountain Development, Khumaltar, Lalitpur, Nepal

Correspondence: Rijan Bhakta Kayastha (rijankayastha@gmail.com)

Published: 18 November 2024

Abstract. In recent decades, the Tsho Rolpa glacial lake has witnessed a significant increase in its surface area. Situated in the Eastern Himalayas at an elevation of 4552 m a.s.l. (meters above sea level), this lake has drawn attention due to the potential risks of glacial lake outburst floods (GLOFs). To comprehensively study and prepare for future GLOF events from Tsho Rolpa, advanced modeling techniques were employed, utilizing the National Weather Service BREACH (NWS-BREACH) and Hydrologic Engineering Center's River Analysis System (HEC-RAS) models. The assessment included the evaluation of dam breach scenarios, especially 20 and 40 m breaches, using the NWS-BREACH model to estimate breach hydrographs. This analysis revealed peak flow rates ranging from 8198 to 26 662 $m^3 s^{-1}$ for the respective breaching scenarios. Additionally, considering a change in the lake area over 20 to 40 years, the peak flow ranged from 8226 to 10 662 $m^3 s^{-1}$ for the 20 m breach scenario. The peak flow estimates obtained from NWS-BREACH were subsequently integrated into the HEC-RAS model to simulate peak discharge and flood heights at varying distances downstream of the lake outlet. To understand the consequences of a GLOF, the impact on downstream areas was assessed through flood inundation maps, and precautionary measures are recommended.

1 Introduction

Glaciers are highly sensitive to climate change due to their relatively quick response to temperature variation: climate cooling results in glacier advance, whereas warming leads to glacier retreat. Therefore, glaciers are excellent indicators of climate change, which has recently had a significant impact in high-mountain glacial environments. Hence, recent glacial retreats and concomitant glacial lake formation/expansion in mountain areas serve as an example and infallible evidence of climate change. Glacial wastage or recession, most often of debris-covered glaciers, causes climate warming due to exposed low-albedo surface area, and the resulting formation of glacial lakes and their expansions are intertwined. In other words, glacial wastage initially creates an environment conducive to glacial lake formation and enlargement. In the later stages of glacial wastage, supraglacial and proglacial lakes create conditions that can enhance and accelerate glacier recession. Sometimes, this leads to potent terminus calving into the lake, which is even reported to be a climate-independent glacier ice-loss mechanism, thereby making the glacier terminus increasingly unstable. This process of glacier terminus ice calving into the glacial lake can, in turn, trigger a glacial lake outburst flood (GLOF), among other glacial lake failure mechanisms such as suddenly high discharge from a seepage point, the collapse of a moraine by a huge ice and snow avalanche into the lake, or an earthquake. In light of this situation, understanding the response of glaciers and glacial lakes to rising air temperatures has become an essential aspect of water resource management as well as the management of GLOF disasters.

In 2010, there were 5701 glacial lakes in the Third Pole region, which includes the Pamir–Hindu Kush–Karakoram–Himalayan system and the Tibetan Plateau (Zhang et al., 2015). Landsat 8 Operational Land Imager (OLI) data from 2014 revealed more than 8200 glacial lakes and an overall lake area of more than 830 km^2 (Chen et al., 2017). These glaciers are melting rapidly, leading to the formation of new glacial lakes and the expansion of existing moraine-dammed lakes. Some debris-covered glaciers have supraglacial lakes at their termini that are surrounded and dammed by the moraine formed during the Little Ice Age. Those glacial lakes emerged during recent decades and are now also expanding, mainly by calving at their termini. Understanding the behavior of glaciers and glacial lakes is a vital aspect of GLOF disaster management. Moreover, tracking glacial lake evolution and the study of the morphology around them is important for estimating GLOF danger from the lakes. Mitigation and prevention of GLOF hazards are urgent issues that need to be addressed in the context of water resource development and conservation in the Himalayas. However, the recession of glaciers and the concomitant formation/expansion of glacial lakes may not be synchronous in the entire region. Differences in many influencing factors, such as the climatic variability, local topographic effects, and thickness and spatial distribution of debris cover, lead to differences in the size, number, distribution, and type of glacial lakes as well as disparity in the recessional behavior of glaciers (e.g., faster/slower retreat or even the existence of advancing glaciers). Within the Hindu Kush–Himalayas region, the eastern part contains the highest number of glacial lakes; the lakes in this area are larger in size and the majority are proglacial in nature. In contrast, in the western part of the region, most lakes are supraglacial, less numerous, and smaller. In Nepal, the eastern part also has more and bigger glacial lakes. Tsho Rolpa is the biggest glacial lake in Nepal. The main objective of this study is to estimate the discharge and inundation scenarios downstream if a GLOF occurred under present or future (40 years from 2021) conditions.

2 Study area

Tsho Rolpa glacial lake is located in the Rolwaling Valley, Dolakha District, in the central Nepal Himalayas (27°52′ N, 86°28′ E; 4546 m a.s.l., meters above sea level) and forms the headwaters of Rolwaling Khola, a tributary of the Tama Koshi River (Fig. 1). The lake had been rapidly expanding on the terminal part of the debris-covered Trakarding Glacier from at least 1958 (surface area 0.23 km^2), when the Survey of India performed a topographic survey, to the late 1990s (surface area 1.55 km^2 in 1999), when mitigation work was carried out (Chikita et al., 1999; ICIMOD, 2011). The lake has been considered to be the most potentially dangerous glacial lake in Nepal, especially before engineering work was carried out in 2000 to lower the lake water level and reduce the level of risk with respect to a possible GLOF (ICIMOD, 2011).

3 Data and methods

The flood hydrograph is an important input parameter in GLOF modeling. The other input parameters required are the digital elevation model (DEM), the surface area of the glacial lake, the maximum depth of the lake, and the volume of water in the lake. Furthermore, different characteristics of the dam, such as the top and bottom elevation of the dam, the width of the dam's crest, the dam's maximum height, and the slopes of the dam's faces play an important role in determining the flow hydraulics over the dam. The material properties of the dam affect the rate of structural erosion; therefore, these will also be taken as input parameters (Westoby et al., 2014). The ALOS PALSAR DEM has been used to create a terrain model to set up the modeling and inundation mapping.

3.1 GLOF modeling

The GLOF modeling process includes breaching the moraine dam, its maximum discharge at the time of breaching, the discharge and water level at different locations downstream, and the creation of a flood inundation map for the downstream area. Risk analysis can be done after this information has been established. Table 1 shows the input parameters used in the Tsho Rolpa GLOF modeling. The research design is the same as that shown in Fig. 2 of Maskey et al. (2020). The Froehlich (1995) model is used to estimate the maximum discharge at the time of dam breaching, the National Weather Service BREACH (NWS-BREACH) model is used to breach the dam, and the Hydrologic Engineering Center's River Analysis System (HEC-RAS) 2-D model is used to create the flood inundation map for the downstream area.

3.2 The NWS-BREACH and HEC-RAS models

The NWS-BREACH model has been used to predict the breach characteristics (size, shape, and time of formation) and the breach outflow hydrograph in this study. The estimated peak discharge from the NWS-BREACH model is verified with the Froehlich model (Froehlich, 1995). Very similar values were found; hence, the peak discharge from the NWS-BREACH model is adopted in this study. It was assumed that the flow reaches its peak discharge following a linear rise before decreasing over the same period of time. The parametric model uses empirical observations of previous dam failures, such as the breach width–depth relation, the time of breach formation, and the breach depth, to develop the outflow hydrograph.

However, observations of such parameters are not existent for moraine dams, rendering the parametric approach infeasible. In this regard, the breach erosion model has some advantages over the parametric breach model, as the critical properties used by the model are measurable or can be estimated

Figure 1. Location map of the study area showing the Tsho Rolpa glacial lake in the Dolakha District, Nepal. A map of Nepal is shown in the inset. Publisher's remark: please note that the above figure contains disputed territories.

Table 1. Input parameters used in Tsho Rolpa GLOF modeling.

Parameter	Unit	Tsho Rolpa glacial lake
Moraine dam data		
Surface area of glacial lake	km^2	1.65
Volume of water in glacial lake	Mm^3	87.2
Maximum depth of lake	m	133.5
Top elevation of dam	m a.s.l.	4546
Toe elevation of dam	m a.s.l.	4521
Width of dam crest	m	50
Length of dam crest	m	315
Maximum height of moraine	m	216
Slope of upstream face of dam	$1:z$	1 : 4
Slope of downstream face of dam	$1:z$	1 : 3
Dam material properties		
Grain size (D50*)	mm	10
Porosity	%	30
Cohesiveness (c)	$kN\,m^{-3}$	0
Internal friction angle (ϕ)	degree	35

* A total of 50 % of grains have an average size smaller than 10 mm and 50 % are bigger.

within a reasonable range from a qualitative description of dam materials.

The HEC-RAS 2-D model is used in modeling GLOF inundation. The details of the NWS-BREACH and HEC-RAS models are adopted from Maskey et al. (2020) in this study.

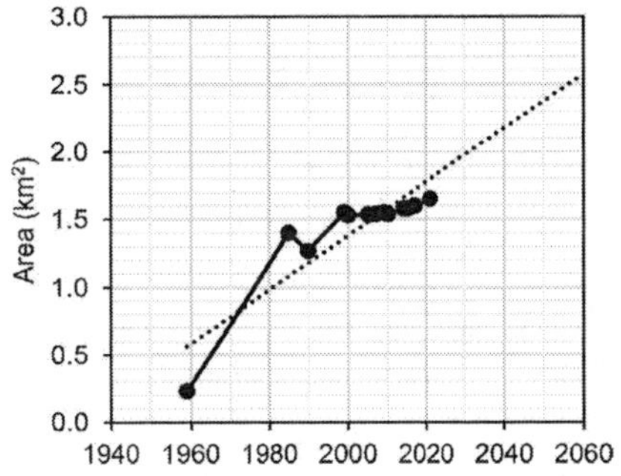

Figure 2. Variation in the area of the Tsho Rolpa glacial lake from 1959 to 2021 and beyond. The dotted line shows a trend line.

4 Results and discussion

Based on available data for the Tsho Rolpa glacial lake from 1959 to 2021 (from ICIMOD until 2011 and from Google Earth satellite images thereafter until 2021), its projected area in 2040 and 2050 is shown in Fig. 2. A linear expansion of the lake area is considered. With these areas in 2040 and 2050, the volume of water in the lake is estimated, and GLOF modeling is carried out. The area of the lake was 0.23 and 1.65 km^2 in 1959 and 2021, respectively, based on previous reports. Using linear expansion, the area of the lake will be 2.33 km^2 in 30 years and 2.53 km^2 in 40 years.

The possible breaching time and peak floods are estimated for breaching scenarios of 20 and 40 m using the NWS-BREACH model. The estimated peak discharges are

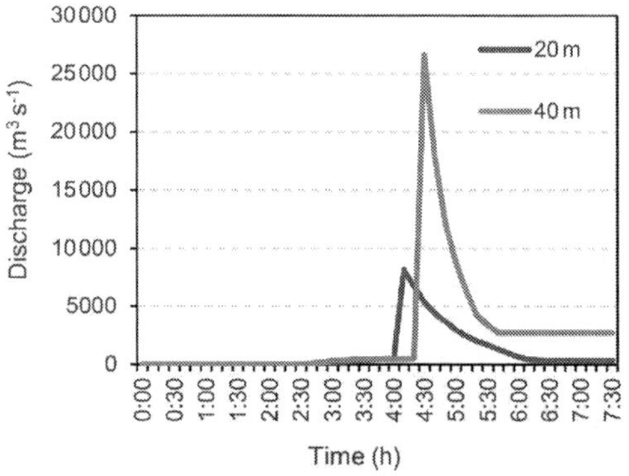

Figure 3. Variation in the peak discharge from the NWS-BREACH model for a respective 20 and 40 m breach in the Tsho Rolpa glacial lake moraine.

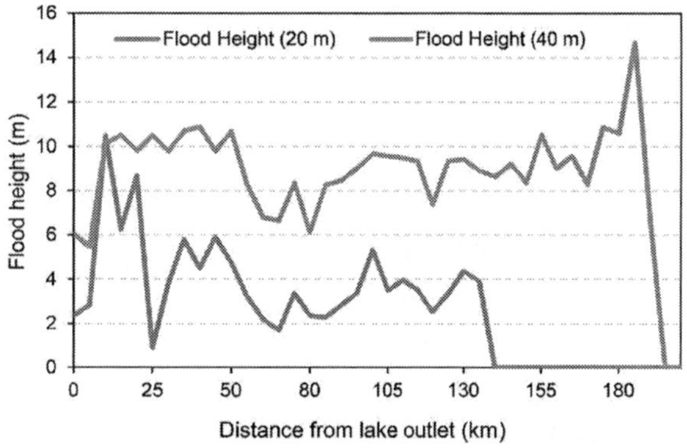

Figure 4. Distribution of the flood water height for a respective 20 and 40 m breach in the Tsho Rolpa glacial lake moraine.

$8198\,m^3\,s^{-1}$ in 4 h 10 min and $26\,662\,m^3\,s^{-1}$ in 4 h 30 min for the 20 and 40 m breaching scenarios, respectively, as shown in Fig. 3. Similarly, the maximum estimated flood heights are 11.4 m at 20 km downstream of the lake and 14.7 m at 185 km downstream of the lake for the 20 and 40 m breaching scenarios, respectively, as shown in Fig. 4. The peak discharge due to the 20 m breach was comparable with the GLOF modeling of ICIMOD (2011), in which the peak discharge was estimated to be $7242\,m^3\,s^{-1}$; this is a reasonable value with respect to the lower surface area and volume of water in 2009 compared with that in 2021. However, the timing of peak discharge in the present study is a little too high, possibly due to the high volume of water at present. Moreover, the flood height is lower than the aforementioned previous study, possibly due to changes in the river morphology at present. The peak flood height due to a 40 m moraine breach at around 185 km downstream of the lake is mainly due to the river cross section.

The inundation map due to a Tsho Rolpa GLOF for a 20 m breach in the 2021 lake scenario generated using HEC-RAS

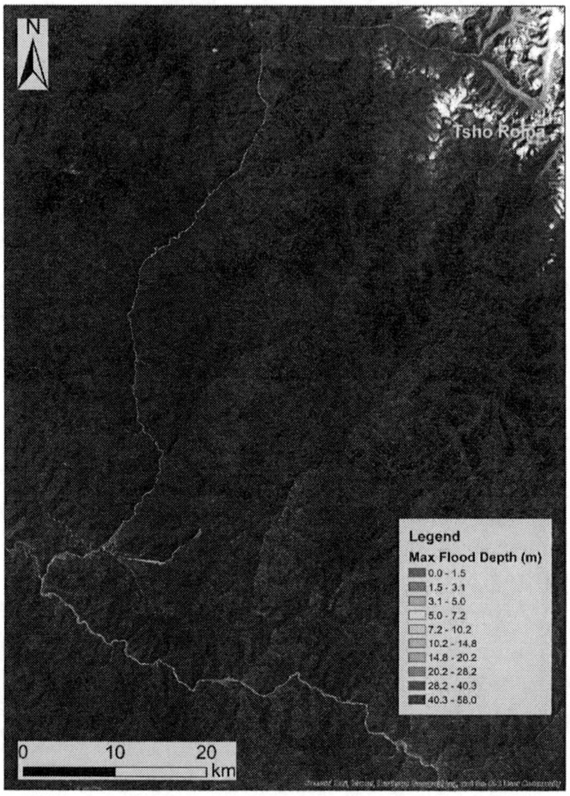

Figure 5. Inundation map of Tsho Rolpa GLOF for a 20 m breach in the 2021 lake scenario. Maximum flood depth is shown in the legend.

is shown in Fig. 5. The maximum flood height is only 1.5 m. The NWS-BREACH modeling is also carried out to estimate the peak discharges for a 20 m breach in 20, 30, and 40 years (from 2021) to check the future GLOF scenarios. The estimated peak discharge for a 20 m breach scenario in 40 years is $10\,662\,m^3\,s^{-1}$, as shown in Fig. 6. The estimated peak discharge in 40 years from the 2021 scenario is increased by $2464\,m^3\,s^{-1}$, which seems to be reasonable compared to its $0.88\,km^2$ area increase.

5 Conclusion

GLOF modeling of the Tsho Rolpa glacial lake has been carried out successfully and has provided estimates of the peak discharges in the 2021 lake scenario and for 20, 30, and 40 years from 2021. In the extreme case, in which a 40 m lake outlet is breached, the estimated peak discharge is $26\,662\,m^3\,s^{-1}$ 185 km downstream in Sunkoshi River, Udayapur District. This point is about 25 km downstream of the confluence of the Dudh Koshi River in Sunkoshi; hence, a GLOF in the Dudh Koshi River basin can cause a com-

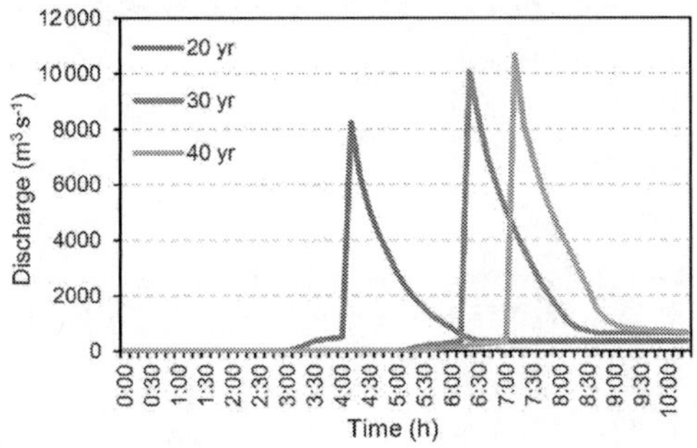

Figure 6. Peak discharges in 20, 30, and 40 years (from 2021) derived from the NWS-BREACH model for a 20 m breach in the Tsho Rolpa glacial lake moraine.

pounding effect if an outburst flood occurs at the same time. At that time, the estimated maximum flood water height is 14.66 m in the Sunkoshi River. Although a GLOF due to the 20 and 40 m breaching scenarios showed a sufficient lead time for early warning and human evacuations, mainly due to a long stretch of the river, there might be ponding during water transit due to possible landslides or debris carried by the flood water. Therefore, precautionary measures, such as community-based GLOF early-warning systems and mechanisms allowing close observation in the case of GLOF events, should be established in GLOF-prone regions.

Data availability. The ALOS PALSAR RTC DEM from 2015 is used in this study. The dataset was developed by the Japan Aerospace Exploration Agency (JAXA) and other collaborators. We downloaded the data on 6 April 2024 from https://asf.alaska.edu/datasets/daac/alos-palsar-radiometric-terrain-correction/ (JAXA et al., 2024).

Author contributions. RBK developed the main idea and wrote the manuscript. SM collected and analyzed the data.

Competing interests. At least one of the (co-)authors is a guest member of the editorial board of *Proceedings of IAHS* for the special issue "Mountain Hydrology and Cryosphere". The peer-review process was guided by an independent editor, and the authors also have no other competing interests to declare.

Disclaimer. Publisher's note: Copernicus Publications remains neutral with regard to jurisdictional claims made in the text, published maps, institutional affiliations, or any other geographical representation in this paper. While Copernicus Publications makes every effort to include appropriate place names, the final responsibility lies with the authors.

Special issue statement. This article is part of the special issue "Mountain Hydrology and Cryosphere". It is a result of the International Conference on Mountain Hydrology and Cryosphere, Kathmandu and Dhulikhel, Nepal, 9–10 November 2023.

References

Chen, F., Zhang, M., Tian, B., and Li, Z.: Extraction of glacial Lake outlines in Tibet plateau using Landsat 8 imagery and google earth engine, IEEE J. Sel. Top. Appl., 10, 4002–4009, 2017.

Chikita, K., Jha, J., and Yamada, T.: Hydrodynamics of a supraglacial lake and its effect on the basin expansion: Tsho Rolpa, Rolwaling Nepal Himalaya, Arct. Antarct. Alp. Res., 3, 58–70, 1999.

Froehlich, D. C.: Peak outflow from breached embankment dam, J. Water Res. Pl., 121, 90–97, 1995.

ICIMOD: Glacial lakes and glacial lake outburst floods in Nepal, Kathmandu, ICIMOD, 99 pp., 2011.

Japan Aerospace Exploration Agency (JAXA) and other collaborators: ALOS PALSAR RTC DEM, ASF [data set], https://asf.alaska.edu/datasets/daac/alos-palsar-radiometric-terrain-correction/, last access: 6 April 2024.

Maskey, S., Kayastha, R. B., and Kayastha, R.: Glacial Lakes outburst floods (GLOFs) modelling of Thulagi and lower Barun Glacial Lakes of Nepalese Himalaya, Progress in Disaster Science, 7, 100106, https://doi.org/10.1016/j.pdisas.2020.100106, 2020.

Westoby, M. J., Glasser, N. F., Brasington, J., Hambrey, M. J., Quincey, D. J., and Reynolds, J. M.: Modelling outburst floods from moraine-dammed glacial lakes, Earth Sci. Rev., 134, 137–159, https://doi.org/10.1016/j.earscirev.2014.03.009, 2014.

Zhang, G., Yao, T., Xie, H., Wang, W., and Yang, W.: An inventory of glacial lakes in the Third Pole region and their changes in response to global warming, Global Planet. Change, 131, 148–157, 2015.

Proc. IAHS, 387, 65–71, 2024
https://doi.org/10.5194/piahs-387-65-2024

Open Access PIAHS

Backwater effect in lowland regions due to bridge structure: a case study of Shreekhandapur, Kavre, Nepal

Rizbi Buddhacharya, Sailesh Maharjan, Rupesh Choudhary, Shyam Sundar Khadka, and Santosh Chaudhary

Department of Civil Engineering, School of Engineering, Kathmandu University, 6250, Dhulikhel, Nepal

Correspondence: Shyam Sundar Khadka (sskhadka@ku.edu.np)

Published: 18 November 2024

Abstract. The backwater effect can be caused by a number of factors, including the bridge's pier width and deck level. In lowland regions, the backwater effect can be particularly problematic, as it can lead to flooding and inundation of farmlands and houses. This study examines the impact of bridge structure on the backwater effect in the lowland region of Shreekhandapur, Kavre, Nepal. The study area is characterized by a hill slope and inner river valley, with a gentle river profile that results in low flow velocity and increased deposition. The study compares the backwater effect of an old demolished bridge with four piers of 1.2 m width and abutments on both sides to a new bridge with a single pier of 0.4 m width and a deck level that is 1 m higher. HEC-RAS software was used to analyse the steady flow and create a flood hazard map. The results show that the new bridge with a reduced pier width and increased deck level significantly reduces the backwater effect. The flood hazard map shows that the new bridge along with floodwalls reduces the potential hazard-prone areas by up to 50 %. The findings of this study can be applied to disaster mitigation and bridge design in lowland regions. By reducing pier width and increasing deck level, bridges can be designed to minimize the backwater effect and reduce the risk of flooding.

1 Introduction

Bridges are remarkable engineering structures that play a crucial role in connecting two separate landmasses or crossing obstacles such as rivers, valleys, or other physical barriers. These bridges have been a part of human history for thousands of years, evolving from simple wooden or stone structures to intricate, modern designs made of steel, concrete, and other advanced materials.

Flooding is the most common natural catastrophe in Nepal. In addition to river flooding during the monsoon, other flood hazards include flash floods caused by excessive rainfall in high areas, glacial lake outburst floods (GLOFs), landslide-induced flooding, and infrastructure flooding (such as embankment failure). Nepal is considered to be the second-highest South Asian nation vulnerable to flood risks. Floods in Nepal between 1954 and 2018 resulted in 7599 fatalities, 6.1 million impacted, and USD 10.6 billion in losses to the economy (EmDAT, 2019). In 2017, 179 people died, more than 75 people were injured, 42 individuals went missing, and the impact extended to 15 303 households (Sharma et al., 2019). Flash flooding can happen in hilly and mid-hill areas when heavy rain falls in a brief period of time, causing high levels of surface water to flow into existing streams and rivers.

Backwater means the increase in water surface level immediately upstream from bridges caused by the resulting reduction in the conveyance area. In a study of the causes of failure of 143 bridges worldwide, it was found that 70 % failed due to flood events (Shamkhi and Salim, 2018). This highlights the importance of understanding the interaction of water flow with bridge structures, rather than considering the bridge structure in isolation.

Therefore, understanding the backwater effect and its implications is crucial for the selection of safe and efficient design of bridges.

Figure 1. Punyamata Khola urbanized area, 2002 (source: © Google Earth).

Figure 2. Punyamata Khola urbanized area, 2023 (source: © Google Earth).

Urban flooding is brought about by heavy rain falling on built-up surfaces with limited drain capacity. Urban flooding can be summed up as "excessive runoff in developed urban areas, where the storm water has nowhere to go due to low drainage system capacity, creating inundations."

An increase in the frequency of urban flood events is becoming a global phenomenon, posing a significant challenge to planners worldwide. The size of these flood events can vary, ranging from a small neighbourhood occurrence to a major catastrophe that submerges urban areas for several hours to multiple days.

Urban floods are a non-natural disaster brought on by overworked drainage systems, haphazard construction, and a disrespect for hydrogeomorphology and natural terrain. Flooding in urban areas is induced not only by unusually heavy rains but also by our cities' misguided approach to urbanization (Shrestha, 2022).

Figures 1 and 2 indicate the unplanned haphazard urbanization in which inundation may cause various problems.

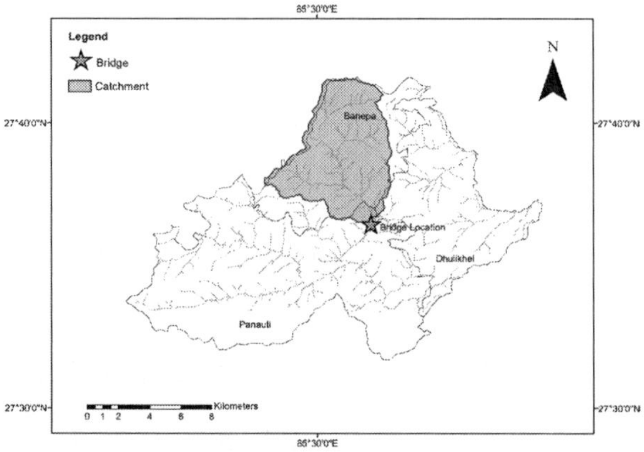

Figure 3. Location of Punyamata Khola bridge and the catchment area.

2 Materials and methods

2.1 Study area

Every year the monsoon flood inundates the study area, i.e. the Shreekhandapur area. The Punyamata Khola bridge is situated across the Punyamata River in Shreekhandapur, Kavre. The study area is located in Panauti Municipality, Ward no. 2, Kavrepalanchok District, in Bagmati Province, Nepal. The study area is characterized by a hill slope and inner river valley, with a gentle river profile of 1 % that results in low flow velocity and increased deposition of soil sediments. Every monsoon, due to heavy rainfall and cloudburst in the Nala area, there is flash flooding of the Punyamata River. Every year, homes and agricultural areas are submerged as a result of flash flooding. This results in loss of life and property. In the Pulbazar region of Banepa, the flooding results in traffic jams, halting the traffic for several hours till the water level resides. It is estimated that approximately NPR 50 lakh[1] financial loss is incurred if the traffic is disrupted for an hour (Shrestha, 2022). Floods also swept away cooking gas cylinders from Sagarmatha Gas Udhyog, Surya Gas Udhyog, and Chandeshwori Gas Udhyog in 2016.

Figure 3 indicates the Punyamata bridge site location and the catchment area.

2.2 Input data

2.2.1 Bridge description

The salient features of the old and new bridge are given in Table 1.

The old Punyamata Khola bridge and the new Punyamata Khola bridge are shown in Figs. 4 and 5, respectively, and their salient features are shown in Table 1. From these salient

[1] 1 lakh is equal to 100 000.

Table 1. Salient features.

	Old bridge	New bridge
Bridge type	Brick masonry with concrete deck slab	Reinforced cement concrete box type bridge
Bridge span	21 m	21 m
Deck level	1441.94 m a.s.l.	1442.94 m a.s.l.
Deck slab width	3.5 m	10 m
No. of piers	Four	One
Pier width	1.2 m	0.4 m

Figure 4. Old Punyamata Khola bridge.

features, we can clearly see that for the new bridge, a design with fewer piers (i.e. one) than the old bridge (i.e. four) was adopted. Also, the pier width was reduced to 0.4 m in the new bridge from 1.2 m in the old bridge.

2.2.2 Survey data

The survey data of the study area were provided by Dhulikhel Municipality at a resolution of 3.5 m. The digital elevation model (DEM) of the study area was generated with help of the survey data.

2.2.3 Hydrologic and hydraulic data

Table 2 gives the catchment area of the Punyamata bridge site location along with the catchment area of Roshi Khola.

Table 3 shows the maximum monthly river discharge for 1975, and Table 4 shows the hydraulic data of the bridge site location.

Figure 6 shows the daily discharge for July 1975.

Figure 5. New Punyamata Khola bridge.

Table 2. Catchment area data.

River	Catchment area (km^2)
Punyamata	41 (as shown in Fig. 3)
Roshi Khola	87

2.3 Methodology

2.3.1 Flood calculation methodology

A DEM generated from survey data was used in HEC-RAS topography, and then data from Table 4 were used to model the bridge in HEC-RAS. Flow data from Table 3 and the method by Betram Darell Richards (hereafter B. D. Richards) were used to estimate the high flood level for the old and new bridge. Using the DEM, cross sections were modelled along the river path. Analysis of flood data was done in four different ways: 1D and 2D analysis, each with the old bridge and the new bridge.

As no stations were available in the catchment area for the precipitation calculation, data from the Roshi Khola catchment area were used to calculate flood discharge as a reference for our site.

For 1D analysis, different discharge levels were measured, as shown in Table 3 and as calculated by B. D. Richards' method. The piers and the wing wall were modelled as shown in Figs. 8 and 9. Ineffective flood areas were also assigned to the left and right side of the bridge. The backwater effect was

Table 3. Maximum monthly flow for 1975.

Month	Discharge (Roshi Khola) $m^3\,s^{-1}$	Discharge (Punyamata Khola) $m^3\,s^{-1}$
January	1.85	0.87
February	1.6	0.75
March	1.31	0.62
April	1.22	0.57
May	1.22	0.57
June	3.83	1.80
July	84.6	39.87
August	28.2	13.29
September	20.6	9.71
October	7.59	3.58
November	2.72	1.28
December	2.1	0.99

Table 4. Hydraulic data.

Design discharge	39.87 $m^3\,s^{-1}$
River bed level (bridge section)	1440.2 m a.s.l.
Flood wall level	1443 m a.s.l.
Contraction coefficient	0.1
Expansion coefficient	0.3
Manning's coefficient	0.03 (natural stream)
River bed slope	0.01

observed for 100 m upstream of the bridge, which showed supercritical flow approaching the bridge.

As the catchment was ungauged, we used the B. D. Richards method to validate the discharge obtained by the catchment area ratio (CAR) method. The meteorological data of four stations near the catchment were recorded to determine the rainfall intensity in the catchment. Using the rainfall data, we determined the flood discharge for a return period of 50 years.

Figure 7 shows the methods used for flood simulation.

2.3.2 Catchment area ratio method

The estimation and the modelling of the availability and quality of water supply and the ecological assessment require reliable estimation of flow. The ratio method is used to estimate flow in an ungauged catchment when a nearby gauged watershed is present for use as a reference. The method estimates flow at an ungauged location by multiplying the measured flow at the nearby reference gauge by the area ratio of the ungauged to gauged watershed. Discharge of an ungauged river is calculated using the following formula:

$$Q_2 = \frac{A_2}{A_1} \times Q_1, \quad (1)$$

where Q_2 is the discharge of the studied (ungauged) river, Q_1 is the discharge of reference (gauged) river, A_2 is the area of the studied catchment, and A_1 is the area of the reference catchment.

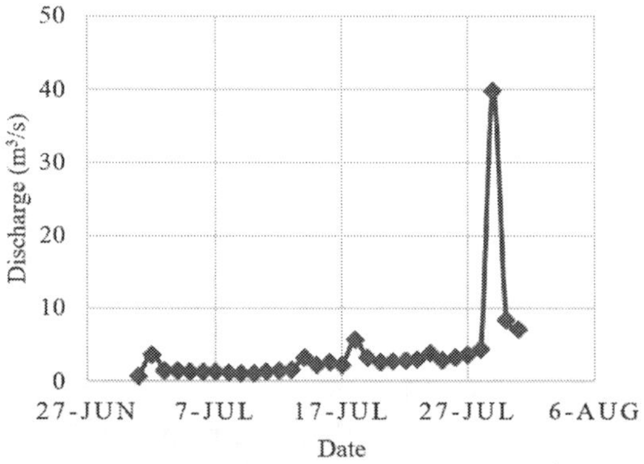

Figure 6. Daily discharge for July 1975.

Table 5. Parameters used in estimation of flood discharge by B. D. Richards' method (Shakya, 2022).

Catchment slope (S)	7	%
Catchment length (L)	10	km
24 h rainfall for return period of 50 years (R_T)	47	$mm\,h^{-1}$

For 2D analysis, flood data from 1 July 1975 to 31 July 1975 were taken as this month recorded the highest rainfall in the region, as shown in Fig. 6. Meshing of the surface was done along the river between the inlet and outlet. For analysis, the bridge and piers were modelled as shown in Figs. 8 and 9.

2.3.3 B. D. Richards' method

B. D. Richards' method may be used for flood estimation using rainfall and basin characteristics. The flood discharge Q in $m^3\,s^{-1}$ is calculated using the following equation:

$$Q = 0.222AIF, \quad (2)$$

where A is the basin area (in km^2), I is the rainfall intensity corresponding to the time of concentration T_C, and F is an aerial reduction factor given as follows:

$$F = 1.09352 - 0.6628\ln(A). \quad (3)$$

The Thiessen polygon method is used to determine the rainfall intensity of the ungauged catchment. The assessment of flood discharge indicates that the peak discharge increases with an increase in duration of the return period. B. D. Richards' method gave the higher flood discharge for a return period of 50 years, and hence it is used in this study to estimate the flood discharge.

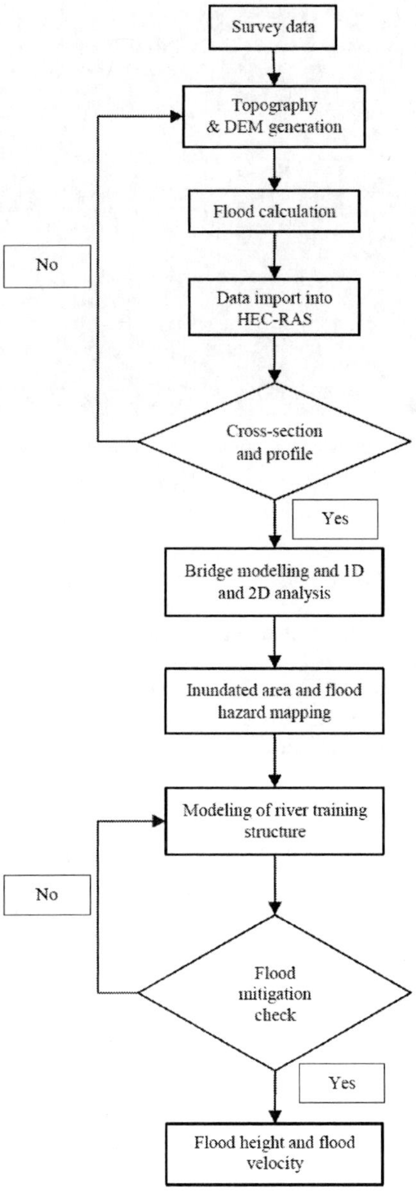

Figure 7. Methodology for flood simulation.

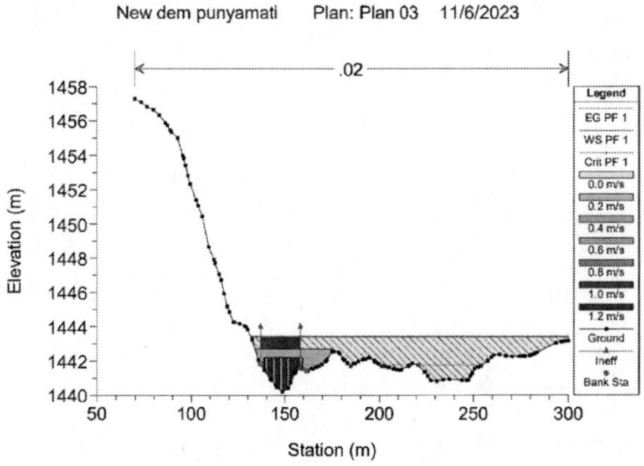

Figure 8. High flood level with the old bridge.

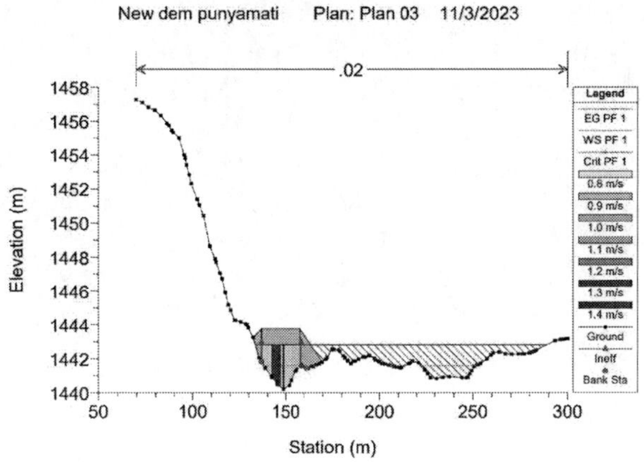

Figure 9. High flood level with the new bridge.

3 Results and discussions

For financial reasons, the bridge is typically situated where the least deck span is present. HEC-RAS was used to model the backwater effect caused by the old bridge. With a high flood level of 1443.46 m a.s.l., it was discovered that the water level would surpass the deck level of the old bridge.

The river discharge calculated by B. D. Richards' method is given in Table 6.

The discharge was calculated to be 47.69 $m^3 s^{-1}$. For the discharge calculation by B. D. Richards, the data from Table 5 have been used.

By simulating a new bridge with a higher deck level and fewer piers, the high flood level was lowered from 1443.46 m a.s.l., as shown in Fig. 8, to 1442.86 m a.s.l., as shown in Fig. 9. The new bridge lowered the flood level by 0.6 m as a result of the backwater effect caused by the previous bridge's piers.

Even though the new bridge passed the maximum flood discharge, there was minimum change in the inundation area, as shown in Fig. 10. Because this study area is a lowland area with a modest slope to the river bed, the bridge's placement created a bottleneck in the terrain, which caused the upstream area to flood.

If not for the piers, supercritical flow would have continued and the flow velocity would have continued to reduce with the new bridge. This can be explained by the reduction

Table 6. Discharge calculation by B. D. Richards' method (Shakya, 2022).

Variable	Value	Units
Time of concentration in hours (T_C)	2.7	$mm\,h^{-1}$
Areal reduction factor (F)	0.85	–
$D = \left(1.102L^2\right)/S \times F$	18.58	–
$R_{T_C} = (R_T \times 22.127 \times T_C^{0.476577})/100$	16.70	–
$I = R_{T_C}/T_C$	6.18	–
$K_R = 0.651(T_C + 1)$	2.41	–
$C_{K_R} = 0.95632/K_R^{1.4806}$	0.26	–
Third estimate of time of concentration (T_{C3})	4.83	$mm\,h^{-1}$
Second estimate of time of concentration (T_{C2})	2.64	$mm\,h^{-1}$
Percent difference	−2 %	
Flood discharge (Q_R)	47.69	$m^3\,s^{-1}$

Figure 10. Flooding of study area without a floodwall (source: © Google Earth).

in piers in the bridge model, which is in agreement with the analysis by Costabile et al. (2015).

Thus, we modelled a floodwall with a height of 3 m upstream of the bridge to lessen the area that would be flooded. The floodwall model led to more than 50 % reduction in the inundation area, as shown in Fig. 11.

4 Conclusions

This study focuses on the backwater effect caused by the bridge's pier width and deck level in lowland regions. The practice of constructing pier bridges should be minimized in low-lying areas as they can be susceptible to flooding. The selection of bridge site should be properly analysed considering both the economic and topographical site conditions. River training works, which include measures such as embankments, spurs, and revetments, can indeed be very effective in preventing flooding and inundation in these regions. So, rather than designing bridges in isolation, environmental factors should be considered as well. Using the 3 m floodwalls in our study helped to reduce the flooding by more than 50 %.

Figure 11. Flooding of study area with a floodwall (source: © Google Earth).

Moreover, urban planning plays a crucial role in mitigating such problems. Future urban planning should consider factors such as flood zoning and proper drainage systems to prevent waterlogging. It is important to ensure that infrastructure is resilient and capable of withstanding extreme weather events.

Data availability. Survey data were provided by Dhulikhel Municipality at a resolution of 3.5 m. The digital elevation model (DEM) of the study area was generated with the help of survey data. The meteorological data were taken from the Department of Hydrology and Meteorology of Nepal and can be provided upon request.

Author contributions. SSK did the conceptualization, methodology, formal analysis, and supervision. SC participated in conceptualization, methodology, and supervision. RB participated in data collection, data analysis, and writing (original draft and review and editing). SM participated in data analysis, interpretation of results, and writing (review and editing). RC participated in data collection and writing (review and editing).

Competing interests. The contact author has declared that none of the authors has any competing interests.

Disclaimer. Publisher's note: Copernicus Publications remains neutral with regard to jurisdictional claims made in the text, published maps, institutional affiliations, or any other geographical representation in this paper. While Copernicus Publications makes every effort to include appropriate place names, the final responsibility lies with the authors.

Special issue statement. This article is part of the special issue "Mountain Hydrology and Cryosphere". It is a result of the International Conference on Mountain Hydrology and Cryosphere, Kathmandu and Dhulikhel, Nepal, 9–10 November 2023.

Acknowledgements. Dipesh Thapa of Zest Engineering Solutions Pvt. Ltd is acknowledged for his guidance on hydraulic modeling.

References

Costabile, P., Macchione, F., Natale, L., and Petaccia, G.: Comparison of scenarios with and without bridges and analysis of backwater effect in 1-D and 2-D river flood modeling, CMES-Comp. Model. Eng., 109–110, 81–103, 2015.

EmDAT: The CRED/OFDA International disaster Database, https://www.emdat.be/ (last access: 3 October 2023), 2019.

Shakya, D. M.: Guidelines on Hydrologic and Hydraulic Analysis and River Training Works for Bridge Design, Department of Roads, https://dor.gov.np/uploads/publication/publication_1642584636.pdf (last access: 13 October 2023), 2022.

Shamkhi, M. S. and Salim, M.: Experimental study of the effect of contraction ratio on backwater by using bridge pier as flow obstruction, 3rd International Energy & Engineering Congress Proceedings Book, https://www.researchgate.net/publication/328723349_EXPERIMENTALL_STUDY_OF_THE_EFFECT_OF_CONTRACTION_RATIO_ON_BACKWATER_BY_USING_BRIDGE_PIER_AS_FLOW_OBSTRUCTION (last access: 13 October 2023), 2018.

Sharma, T. P., Zhang, J., Koju, U. A., Zhang, S., Bai, Y., and Suwal, M. K.: Review of flood disaster studies in Nepal: A remote sensing perspective, Int. J. Disast. Risk Re., 34, 18–27, 2019.

Shrestha, A.: Framing the Challenges of Urban Flooding on Accessibility, in: Proceedings of 12th IOE Graduate Conference, 19–22 October 2022, Thapathali Campus, IOE, TU, Kathmandu, Nepal, http://conference.ioe.edu.np/publications/ioegc12/IOEGC-12-066-12101.pdf (last access: 15 October 2023), 2022.

Proc. IAHS, 387, 73–77, 2024
https://doi.org/10.5194/piahs-387-73-2024

Open Access PIAHS

Augmented-reality-based snow visibility simulation for disaster preparedness in the Western Himalayas

Sanjay Saifi and RAAJ Ramsankaran

Hydro-Remote Sensing Applications (H-RSA) Group, Department of Civil Engineering,
Indian Institute of Technology Bombay, Powai, Mumbai, India

Correspondence: Sanjay Saifi (sanjaysaifi241@gmail.com) and RAAJ Ramsankaran (ramsankaran@civil.iitb.ac.in)

Published: 18 November 2024

Abstract. Snow-induced disasters pose significant risks in mountainous regions. Accurate visibility assessment is crucial for informed decision-making and disaster preparedness. This paper presents an innovative approach that integrates theoretical literature with practical application using augmented reality (AR) to address this challenge. The proposed system, named Him-Drishti, harnesses the established correlation between snowfall intensity and visibility to create a predictive model. By incorporating user-input snowfall intensity predictions, the Him-Drishti AR application dynamically overlays real-time visibility simulations onto the physical environment. Moreover, the use of smartphones further amplifies the practicality of this innovative solution, making the AR application a handy tool, even in the most challenging terrain. This intuitive visualization empowers stakeholders with a comprehensive understanding of potential risks, facilitating informed decisions and optimized disaster response strategies. Through AR visualization, stakeholders, emergency responders, and local authorities gain valuable insights into visibility conditions, enabling them to make informed decisions, take proactive measures, and allocate resources effectively. This study not only enhances the understanding of snow-related disasters but also demonstrates the capacity of AR in disaster management and risk reduction.

1 Introduction

In recent times, there has been an apparent rise in extreme weather and climatic events on a global scale. This trend is anticipated to intensify in terms of both frequency and severity, as emphasized in a report by the Intergovernmental Panel on Climatic Change (IPCC, Climate Change 2023). Although the future prospects for snowfall in midlatitudes appear bleak due to global warming, extreme snowfall occurrences remain a prominent and destructive category of extreme weather phenomenon, especially in the Northern Hemisphere (Quante et al., 2021).

Extreme snowfall events emerge as potent disruptors that are capable of impacting diverse aspects of urban life, particularly in the realms of air and ground transportation (Santiago-Iglesias et al., 2023). Furthermore, snowfall disasters have claimed numerous lives. In a recent occurrence, 27 lives were lost in India in 2022 due to snowfall disasters (Annual Report, 2023). These events emphasize the critical need for advanced tools and technologies to mitigate the impacts of snow-induced disasters.

Adaptive capacity stands as a key component in the realm of disaster risk reduction, as highlighted within the overarching conceptual frameworks of vulnerability and risk (IPCC, Climate Change 2023). This imperative extends beyond immediate hazard response and encompasses climate change impact mitigation (Sharma and Patwardhan, 2008).

A crucial element of adapting to hazards is risk communication. Existing research on risk communication has predominantly focused on the viewpoint of information senders, often neglecting the choices made by receivers (Palttala et al., 2011). There is a noticeable gap in scientific attention when it comes to recipient perspectives on information usage and risk perception during snow events (Adekola et al., 2021).

Conventional methods with respect to visualization techniques, including PowerPoint presentations (Koutitas et al., 2019), didactic training (Rossler et al., 2019), and walkthroughs of systems (Koutitas et al., 2019), have long been utilized. However, recent advancements in extended-reality (XR) technologies have showcased their superiority over these conventional methods. Augmented reality (AR) and virtual reality (VR) have demonstrated enhanced effectiveness in various immersive applications and outperformed their traditional counterparts (Koutitas et al., 2019).

With snow-induced disasters posing significant threats to mountainous areas, there is a critical necessity for innovative solutions in disaster preparedness and mitigation that is exacerbated by climate change. This research addresses the gap in understanding risk perception during extreme snowfall events. Focusing on the hazard of low visibility due to snowfall, the research aims to enhance disaster-response adaptability.

This study introduces an AR-based application named Him-Drishti. The motivation for this study stems from the essential insights provided by Bendickson (2003), who highlighted the correlation between heightened snowfall intensity and reduced visibility. Him-Drishti overlays virtual snow particles onto the real-world environment, offering a controlled, interactive setting for assessing and experiencing snowfall conditions. Him-Drishti provides a unique simulation experience, allowing users to visualize and understand the challenges posed by reduced visibility during snowfall. This innovative approach not only facilitates better preparedness for disaster response but also opens new avenues for training and awareness in the realm of snow-related emergencies.

2 Study site

The Western Himalayas span from Jammu and Kashmir to the western border of Nepal, encompassing a substantial area of around 331 392 km^2. This region constitutes roughly 10 % of India's overall geographical area. Precipitation in the Western Himalayas, occurring as rain and snow, is influenced by western disturbances from October to May and the southwest monsoon from July to September (Kripalani et al., 2003). Climatic conditions in the area vary with altitude. In the higher regions, temperatures range from 15 to 18 °C in summer, while winters are below the freezing point, leading to permanent snow cover above 4800 m (Rani, 2017).

3 Data and methods

The methodology encompasses several key steps to develop an AR-based snow visibility simulation, as shown in Fig. 2. It begins with the development of virtual snow particles, considering their physical properties and color characteristics. The integration of AR components into real-world environments is achieved by using the features of the Vuforia software development kit (SDK). Finally, the Unity framework is employed for compiling the Android application, resulting

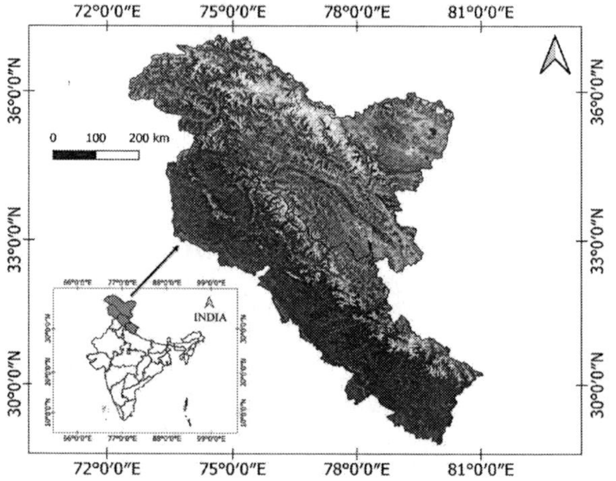

Figure 1. Western Himalayas study region (map data: © Google Maps 2023). Publisher's remark: please note that the above figure contains disputed territories.

in a comprehensive and realistic AR-based snow visibility simulation for smartphones. A detailed methodology is presented in the following sections, providing a comprehensive overview of the steps involved in creating the AR-based snow visibility simulation.

3.1 Snow particle development

In the development of snow particles for the AR-based snow visibility simulation, careful consideration is given to replicating the physical properties of snowfall. To emulate the natural behavior of falling snow, the implementation incorporates realistic gravitational forces that influence the movement of each particle. Snow particle sizes for the various Western Himalayan zones were provided in a study by Negi and Kokhanovsky (2011), and this information is employed in this work to simulate and precisely portray a range of particle sizes, resulting in a visually compelling and true-to-life experience.

Furthermore, careful consideration is given to the color characteristics, transparency, and reflective characteristics of the snow particles in order to improve the simulation's visual accuracy. Snow particles are developed to provide an immersive experience for the user. To incorporate snowfall intensity data into the simulation, the Windy website is utilized as a source of predictive snowfall intensity data. These data are essential to dynamically update the simulation for the user, ensuring that it reflects the predictive snowfall conditions accurately.

3.2 Integration of the AR system into the scene

In this phase, real-world scenes are captured through the smartphone camera to provide the foundation for the simula-

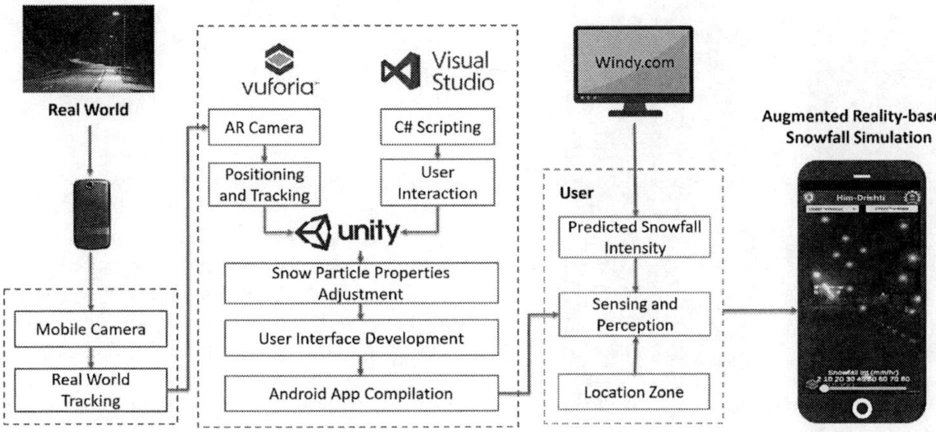

Figure 2. Methodological workflow.

tion. To enhance the scene-capturing capabilities and enable efficient viewing of real-world scenes, the Vuforia SDK is implemented. This SDK not only aids in capturing scenes but also serves as a robust platform for integrating virtual snow particles seamlessly into real-world environments. This integration ensures an immersive and interactive experience for users, allowing them to visualize and interact with the augmented snowfall simulation.

3.3 Graphical user interface (GUI) development

For the development process, Visual Studio is chosen as the integrated development environment (IDE). The development environment plays a crucial role in creating an engaging interface that allows users to interact seamlessly with the AR-based snow visibility simulation.

A snowfall intensity slider is developed to control the simulation with respect to replicating the predictive snowfall intensity conditions from the Windy website. The development of a controlled simulation makes Him-Drishti a valuable tool for users seeking to understand and visualize live snowfall scenarios.

3.4 Compilation of the Android application

Finally, the Unity cross-platform development framework is employed to provide a robust platform for the Android application development. Unity facilitates the creation of a versatile and accessible application that can be deployed across various Android devices. The combination of Unity, the Vuforia SDK, Visual Studio, and predicted snowfall intensity data from the Windy website contributes to the comprehensive methodology for developing an AR-based snow visibility simulation on a smartphone. The tools and techniques utilized for the study are illustrated in Fig. 3. This approach ensures a realistic and engaging user experience while leveraging AR for an innovative application.

Figure 3. Tools and techniques utilized in this study.

4 Results and discussion

The proposed AR-based snow visibility simulation, Him-Drishti, uses the known link between snowfall intensity and visibility to dynamically add virtual visibility simulations on top real-world scenes, as shown in Fig. 4.

The study's innovative integration of theoretical literature with practical application through AR has yielded a promising system for addressing challenges related to accurate visibility assessments of snowfall events. This approach, implemented on smartphones for enhanced practicality, empowers stakeholders with a comprehensive understanding of potential risks, thereby facilitating informed decision-making and optimized disaster response strategies.

Focusing on the Western Himalayas as a study site, the research underscores the critical need for effective disaster-

(a) (b)

Figure 4. **(a)** A real-world snowfall scene **(b)** and a Him-Drishti simulation.

response adaptability and proactive measures in challenging terrain. Him-Drishti, tailored for this region, emerges as a valuable tool for various applications, including weather analysis, emergency response planning, road maintenance, and winter sports management. The correlation between snowfall intensity and visibility provides crucial insights into the relationship between higher snowfall intensity and reduced visibility. The meticulous development of snow particles for the simulation aims to create a visually compelling and immersive experience.

In essence, the results underscore the transformative potential of AR in disaster management and risk reduction, particularly in the context of snow-induced disasters in mountainous regions. This study contributes not only to the advancement of AR applications in disaster management but also to the broader discourse on adapting technology to suit the unique challenges posed by varied geographical and climatic conditions.

5 Conclusions

This study introduces a pioneering AR-based snow visibility simulation system, Him-Drishti, tailored for the Western Himalayas, emphasizing its transformative potential in disaster preparedness and risk reduction. The successful integration of snowfall intensity data into the simulations enables a dynamic and accurate representation of snowfall visibility, providing valuable insights for stakeholders in the region. The correlation between snowfall intensity and visibility, elucidated through the study, contributes nuanced understanding for decision-makers, empowering them to formulate effective disaster risk management strategies. The user-friendly interface enhances adaptability and customization, allowing local authorities and emergency responders to tailor responses based on the specific characteristics of the Western Himalayas.

6 Future scope and limitations

To propel the field forward, future research endeavors should delve into assessing the scalability of the proposed AR snow visibility simulation beyond the Western Himalayas, exploring its applicability in other mountainous regions worldwide. Collaborative efforts with meteorological agencies and disaster management organizations can elevate the practicality and effectiveness of the proposed solution, ensuring its relevance and reliability in varying contexts.

This involves a continual cycle of refining the simulation's accuracy based on user feedback and updated meteorological information. Maintaining a collaborative relationship with stakeholders, including emergency responders and local authorities, is crucial to keeping the application user-friendly and aligned with the evolving needs of these key end users. This iterative approach ensures that the simulation remains a dynamic and responsive tool in the realm of disaster preparedness and risk reduction.

While the proposed AR snow visibility simulation presents a ground-breaking approach, it is essential to acknowledge its current limitations. The simulation does not account for the wind effect, a significant factor during snowstorms that can lead to blizzards. Future iterations could address this limitation by considering wind speed, enhancing the simulation's accuracy, and reflecting a more comprehensive representation of real-world snowfall conditions. Additionally, to further improve the simulation's precision, precise meteorological data can be employed to accurately reproduce snowfall patterns and intensities in the AR environment. Exploring the integration of 360 ° photos for creating simulations in different locations and enhancing the viewing experience with VR and mixed-reality headsets are avenues for future development, ensuring the continuous evolution and advancement of the proposed AR snow visibility simulation system.

7 Availability of the tool: Him-Drishti

The authors are enthusiastic about sharing this innovative AR solution and invite interested readers to contact the authors for access to the tool.

Code availability. The underlying software code is not publicly accessible because the study area might contain international boundaries and disputed areas. However, the authors are enthusiastic about sharing this innovative AR solution and invite interested readers to contact them for access to the code.

Data availability. The underlying research data regarding snowfall intensity prediction are available at the Windy website: https://www.windy.com/-Menu/menu?ptype,21.678,86.880,5 (Windy, 2023). Additionally, information on snow particle sizes for various Western Himalayan zones is provided in the study by Negi and Kokhanovsky (2011, https://doi.org/10.5194/tc-5-831-2011). Finally, the Vuforia software development kit can be accessed at the Vuforia Developer Portal: https://developer.vuforia.com/library/getting-started/vuforia-engine-package-unity (Vuforia Developer Portal, 2023).

Author contributions. SS was responsible for the conceptualization, methodology, formal analysis, and writing – original draft. RR participated in the conceptualization, methodology, supervision, and writing – review and editing.

Competing interests. The contact author has declared that neither of the authors has any competing interests.

Special issue statement. This article is part of the special issue "Mountain Hydrology and Cryosphere". It is a result of the International Conference on Mountain Hydrology and Cryosphere, Kathmandu and Dhulikhel, Nepal, 9–10 November 2023.

References

Adekola, J., Renaud, F. G., and Hill, C.: Risk Information Sources for snow disaster risk preparedness in Scotland, Int. J. Disaster Risk Sci., 12, 854–866, https://doi.org/10.1007/s13753-021-00386-y, 2021.

Annual Report: https://mausam.imd.gov.in/, India Meteorological Department, India Meteorological Department Ministry of Earth Sciences (MoES), Govt. of India, retrieved from https://mausam.imd.gov.in/imd_latest/contents/ar2022.pdf (last access: 14 September 2023), 2023.

Bendickson, S.: Relationship between Visibility and Snowfall Intensity, TP 14151E, https://trid.trb.org/view/755982, retrieved from https://skybrary.aero/sites/default/files/bookshelf/943.pdf (last access: 12 September 2023), 2003.

IPCC, Climate Change 2022: Impacts, adaptation and vulnerability, Intergovernmental Panel on Climate Change (IPCC), https://doi.org/10.1017/9781009325844, 2023.

Koutitas, G., Smith, K., Lawrence, G., Metsis, V., Stamper, C., Trahan, M. H., and Lehr, T.: A virtual and augmented reality platform for the training of first responders of the ambulance bus, ACM Int. Conf. Proceeding Ser, 299–302, https://doi.org/10.1145/3316782.3321542, 2019.

Kripalani, R. H., Kulkarni, A., and Sabade, S. S.: Western Himalayan snow cover and Indian monsoon rainfall: A re-examination with INSAT and NCEP/NCAR data, Theor. Appl. Climatol., 74, 1–18. https://doi.org/10.1007/s00704-002-0699-z, 2003.

Negi, H. S. and Kokhanovsky, A.: Retrieval of snow grain size and albedo of western Himalayan snow cover using satellite data, The Cryosphere, 5, 831–847, https://doi.org/10.5194/tc-5-831-2011, 2011.

Palttala, P., Boano, C., Lund, R., and Vos, M.: Communication Gaps in Disaster Management: Perceptions by Experts from Governmental and Non-Governmental Organizations, J. Contingencies Crisis Manag., 20, 2–12, https://doi.org/10.1111/j.1468-5973.2011.00656.x, 2011.

Quante, L., Willner, S., Middelanis, R., and Levermann, A.: Regions of intensification of extreme snowfall under future warming, Sci. Rep., 11, 16621, https://doi.org/10.1038/s41598-021-95979-4, 2021.

Rani, S.: Evaluating snow cover changing trends of the Western Indian Himalaya, Spat. Inf. Res., 26, 103–112, https://doi.org/10.1007/s41324-017-0158-7, 2017.

Rossler, K. L., Sankaranarayanan, G., and Duvall, A.: Acquisition of fire safety knowledge and skills with virtual reality simulation, Nurse Educ., 44, 88–92, https://doi.org/10.1097/nne.0000000000000551, 2019.

Sharma, U. and Patwardhan, A.: Methodology for identifying vulnerability hotspots to tropical cyclone hazard in India, Mitig. Adapt. Strat. Gl., 13, 763, https://doi.org/10.1007/s11027-008-9145-6, 2008.

Santiago-Iglesias, E., Carpio-Pinedo, J., Sun, W., and Palomares, J. C. G.: Frozen city: Analysing the disruption and resilience of urban activities during a heavy snowfall event using Google Popular Times, Urban Clim., 51, 101644, https://doi.org/10.1016/j.uclim.2023.101644, 2023.

Vuforia Developer Portal: Vuforia software development kit, https://developer.vuforia.com/library/getting-started/vuforia-engine-package-unity, last access: 2 September 2023.

Windy: Snowfall intensity prediction data, https://www.windy.com/-Menu/menu?ptype,21.678,86.880,5, last access: 5 September 2023.

Proc. IAHS, 387, 79–86, 2024
https://doi.org/10.5194/piahs-387-79-2024

Landslide hazard mapping of Wayanad District of Kerala, India, incorporating copula-based estimation of joint probability of rainfall

Shamla Dilama Shamsudeen and Adarsh Sankaran

Department of Civil Engineering, TKM College of Engineering Kollam-05,
APJ Abdul Kalam Technological University, Kerala, India

Correspondence: Shamla Dilama Shamsudeen (dsshamla@gmail.com)

Published: 18 November 2024

Abstract. The development and integration of the spatial and temporal probabilities of landslides are required for complete landslide hazard mapping at any location. Under changing climate, the computation of the temporal probability of landslides with rainfall magnitude alone is inaccurate. This research proposes a framework based on copula functions to develop a landslide probability map using multi-site rainfall data by accounting for the rainfall variables of intensity and duration using a joint-probability approach. The proposed technique is used for Wayanad District, Kerala, India, considering extreme rainfall events in 2018. Firstly, the landslide susceptibility map of the district was developed using a robust random forest (RF) model. Based on regional geology, geomorphology, and climate, different regions of Wayanad have varying rainfall thresholds assessed according to the intensity and duration of the rainfall. Then, the temporal probability of landslides was developed, accounting for the intensity and duration of rainfall events using the joint-probability estimation using copula. Through the integration of the landslide spatial probability map with the temporal probability, landslide hazard maps (LHMs) for Wayanad were developed for time periods ranging from 1 to 50 years. The results of the study indicate the need for bi- or multi-variate landslide probability modeling in studies on regional landslide hazard assessments.

1 Introduction

Landslides, defined as the downward movement of rock, debris, or earth as induced by gravity, have had a profound impact on property and human life. Landslides can be induced by a variety of external factors, including heavy rainfall, earthquakes, changes in water level, storm waves, or human activity, with rainfall being the most common and most major triggering element. Landslide threats are anticipated to increase globally as weather extremes intensify (Crozier, 2010). Landslides have become a common monsoon disaster in Kerala's Western Ghats region in recent years (Ajin et al., 2022; Hao et al., 2020). The distinct physiography and climate of the Western Ghats region are considered to be responsible for the catastrophic landslides that occur along their windward slope during periods of intense rainfall (Achu et al., 2021). The coupling influence of rainfall processes and environmental parameters such as topography, landform, lithology, and fault in the prediction of regional landslides has only been examined in a few studies. Regional landslide hazards have been assessed using probability models. In general, a univariate approach has been employed to examine the impact of climatological variables on landslides. Climate variables are highly correlated, and univariate analysis cannot capture the complex and latent characteristics of climatological events. Thus, multivariate analysis is required to describe the structure of variable dependence. Copula functions enable a multivariate approach in which the dependence structure among random variables is determined irrespective of the marginal distributions.

The main focus of this research is to develop a landslide hazard map for Wayanad District of Kerala, India, using copula functions. The copula functions offer a multivariate approach in this study, where rainfall duration and intensity can be evaluated. The specific objectives are (i) to determine the intensity–duration (ID) relationship between

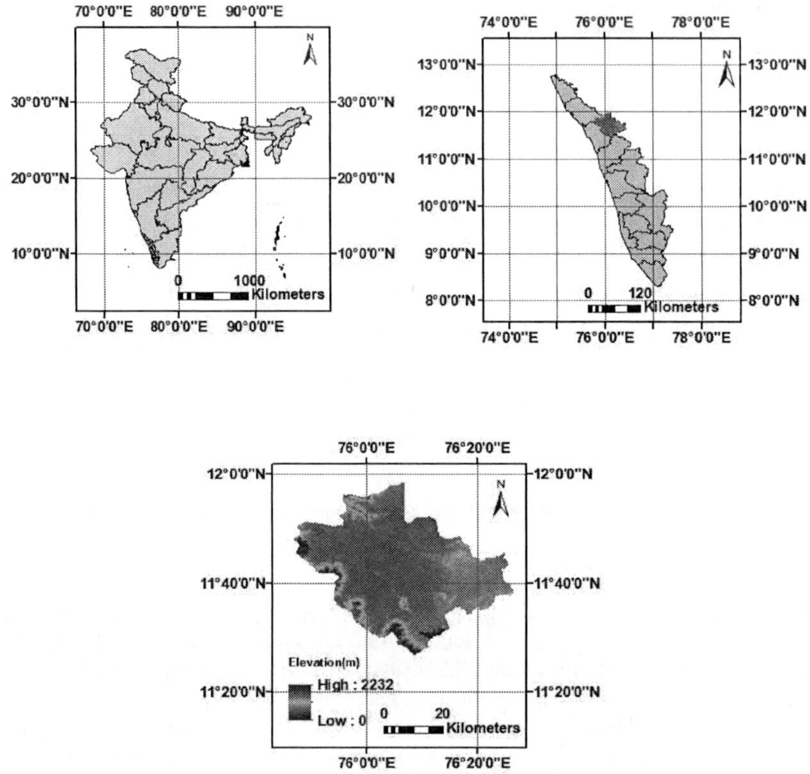

Figure 1. Location and digital elevation model of Wayanad.

landslide occurrence and rainfall, (ii) to prepare the landslide susceptibility map of Wayanad District of Kerala using the machine learning technique of random forests, (iii) to determine the temporal probability of landslides considering multi-site rainfall information using the copula function, and (iv) to prepare the hazard map by integrating the spatial and the temporal probability of landslides.

2 Data and study area

The research was carried out in Wayanad District (Fig. 1) of Kerala, India, which is located in the northeastern part of the state (11–12° N, 75.5–76.25° E). The study region has a total area of 2132 km^2, of which 885.92 km^2 is covered by forest. It is located between 700 and 2300 m a.m.s.l. (above mean sea level).

The average annual precipitation in the region is 2322 mm.

The landslide inventory (Fig. 2) of Wayanad was created for the years 2007 to 2018 using data from the Geological Survey of India's Bhukosh (https://gsi.gov.in, last access: 5 June 2023).

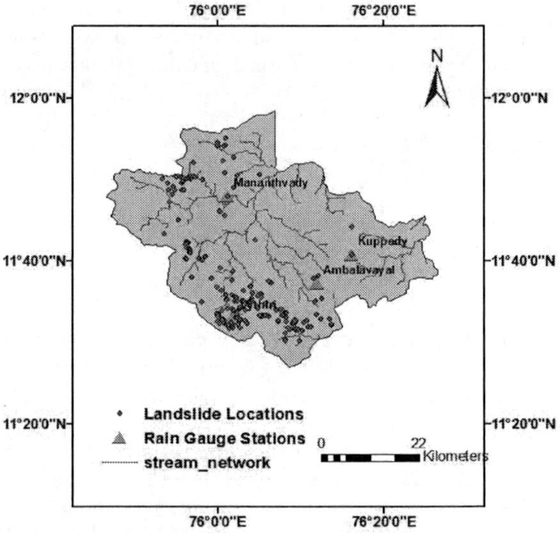

Figure 2. Stream network of Wayanad overlaid by landslide inventory and rain gauge locations.

The study area has been divided into four zones (Fig. 2) based on the proximity to rain gauge stations in order to account for the varying properties of rainfall belonging to the different regions of Wayanad. The rainfall data for the study were obtained from IMD (India Meteorological Department, https://imdpune.gov.in, last access: 5 June 2023), specifically gridded data of $0.25° \times 0.25°$ resolution from 1991 to 2021 at a daily scale. The data were retrieved for four rain gauge stations (Fig. 2): Mananthavady (11.8°, 76.02°), Vythiri (11.5°, 76.03°), Ambalavayal (11.62°, 76.2°), and Kuppady (11.68°, 76.27°).

3 Methodology

This research suggests a copula-based approach that takes the duration and intensity from multi-site rainfall data into account while creating the landslide probability map.

The landslide probability map was prepared by integrating the landslide susceptibility and the temporal probability of the landslide. The susceptibility map was created utilizing the conditioning factors, which included topographical, hydrological, geological, and land use or land cover maps, as well as the landslide inventory map. The joint-probability approach of copula functions was used to obtain the temporal probability of a landslide by calculating the exceedance probability of rainfall parameters, such as intensity and duration, which are considered to be triggering factors of landslide occurrence in the study area. The overall methodology is depicted in Fig. 3.

The rainfall data were collected on a daily basis for different regions of Wayanad. The data were used to determine the rainfall threshold by cross-referencing it with the landslide inventory data from the period of 2007–2018. The empirical rainfall threshold using a power law was calculated using the relationship between rainfall intensity and duration, as well as the landslide events. This was expressed by the following formula:

$$I = \alpha D^{\beta}, \tag{1}$$

where α and β are the intercept and slope parameters, respectively (Segoni et al., 2018; Berti et al., 2020). The rainfall thresholds in Wayanad varied depending on the local, geological, geomorphological, and climatic conditions.

In order to determine the temporal probability of landslides, the rainfall parameter, such as intensity, was acquired by disaggregating the annual maximum rainfall into durations of 3, 5, 7, 15, and 30 d. For this study, we used rainfall data spanning 31 years (1991–2021) to determine the temporal probability of landslides.

The rainfall variables of intensity (I) and duration (D) are modeled using the log-normal distributions and gamma distributions. The parameters were determined using the maximum-likelihood method. The multivariate probability distribution was joined with the univariate marginal distribution functions using the Frank copula, which belongs to the Archimedean family of copulas, C:

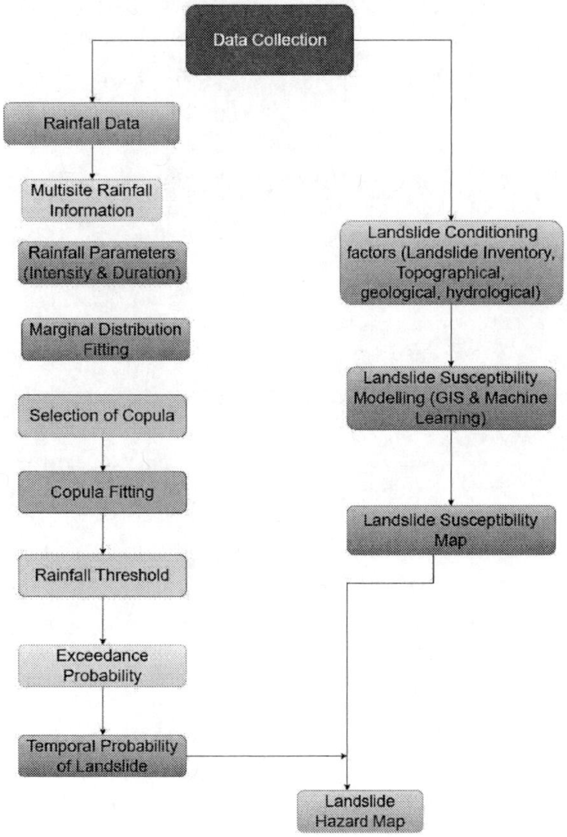

Figure 3. Flowchart showing the overall methodology.

$$C(u,v) = -\frac{1}{\theta}\ln\left[1 + \frac{\left(e^{-\theta u} - 1\right)\left(e^{-\theta v} - 1\right)}{\left(e^{-\theta u} - 1\right)}\right], \tag{2}$$

where u and v are marginal distribution functions of intensity and duration, and θ is the Frank copula parameter. The maximum-likelihood estimation (MLE) approach was also used to estimate the parameters of the copula functions. The dependencies among the rainfall parameters of intensity and duration were determined in terms of Kendall's tau, τ, the correlation coefficient.

The joint exceedance probabilities of rainfall intensity and duration were determined by the theory of the return period (T) using the copula functions. The return period can be defined using the copula theory according to the event for X given $Y \geq y$ or the event for Y given $X \geq x$; these are known as the conditional return period for X given $Y \geq y$ and the conditional return period for Y given $X \geq x$ and are calculated as follows (Shiau et al., 2006; Bezak et al., 2016; Haile, 2022):

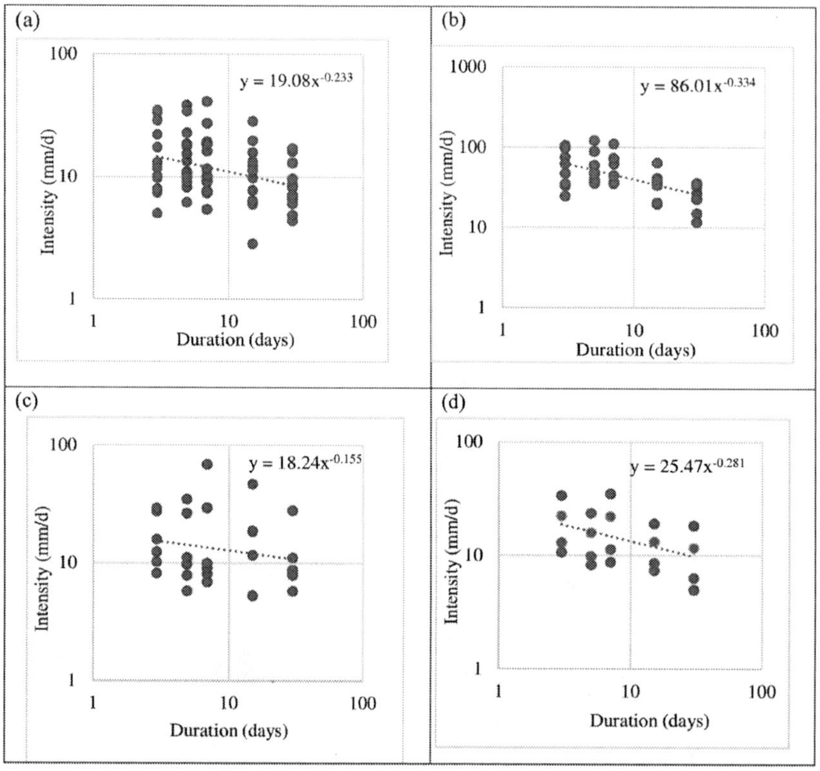

Figure 4. Rainfall threshold obtained at **(a)** Mananthavady, **(b)** Vythiri, **(c)** Ambalavayal, and **(d)** Kuppady.

$$T_{X/Y\geq y} = \frac{1}{[1-F(y)][1-F(x)-F(y)+C[F(x)F(y)]]}, \quad (3)$$

$$T_{Y/X\geq x} = \frac{1}{[1-F(x)][1-F(x)-F(y)+C[F(x)F(y)]]}, \quad (4)$$

where T is the return period, and $F(x)$ and $F(y)$ are cumulative distributions (CDFs) of marginal distributions.

Using the exceedance conditional probability p, the bivariate risk during a period of N years, Pr (N), can be determined as follows (Pabaghi et al., 2023):

$$\Pr(N) = 1-(1-p)^N. \quad (5)$$

Thus, the temporal probability of landslides over N years can be determined using Eq. (5).

The identification of topographical, geological, environmental, and anthropogenic factors that influence the occurrence of landslides in the study area is an important step in landslide hazard mapping. Thus, the following conditioning factors were used in the landslide susceptibility analysis: slope, aspect, curvature, elevation, stream power index (SPI), topographic wetness index (TWI), land use and land cover (LULC), geology, and normalized difference vegetation index (NDVI). The factor maps were created with a 30 m × 30 m resolution SRTM (Shuttle Radar Topography Mission) digital elevation model. The random forest machine learning technique was used for feature selection and landslide susceptibility mapping. The receiver operating characteristic (ROC) curve and the area under the ROC curve (AUC) were used to validate predictions. The landslide hazard maps (LHMs) for different time periods were then created by integrating the landslide susceptibility map (LSM) with the temporal probability of landslide (Lee et al., 2020; Kim et al., 2020).

4 Results and discussion

The hazard mapping under the estimated rainfall threshold was done by integrating the landslide susceptibility mapping and the temporal probability of a landslide occurrence obtained by the copula function.

4.1 Rainfall threshold

The rainfall intensity–duration thresholds (Fig. 4) obtained at Mananthavady, Vythiri, Ambalavayal, and Kuppady are $19.08D^{-0.223}$, $86.01D^{-0.334}$, $18.24D^{-0.155}$, and $25.47D^{-0.281}$, respectively, where D is the duration in days.

Table 1. Copula fitting and parameters.

Station	Copula	Parameter	Kendall's τ
Mananthavady	Frank	−12.5	−0.72
Vythiri	Frank	−9.8	−0.66
Ambalavayal	Frank	−6.34	−0.531
Kuppady	Frank	−8.65	−0.625

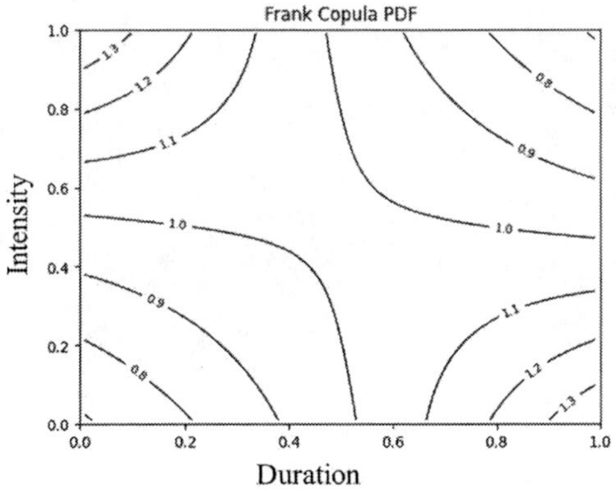

Figure 5. Contour plot of joint PDF of Mananthavady region.

The rainfall threshold values obtained in this study are comparable to those obtained in Wayanad in a previous study by Abraham et al. (2020). The rainfall threshold obtained at Vythiri could be validated with the landslide occurrence on 8 August 2019, which happened with a daily rainfall amount of 192.22 mm and a 3 d rainfall intensity of 81.6 mm d^{-1}.

4.2 Temporal probability assessment of landslides

Using the copula functions, the combined conditional exceedance probability of rainfall intensity and duration is used to calculate the temporal probability of landslides. Table 1 shows the parameters and correlation coefficient that were determined for the Frank copula.

The joint probability distribution function (PDF) (Fig. 5) demonstrates zero tail dependency, indicating that the variables are asymptotically independent of their tails. The negative value of Kendall's τ shows that the rainfall variables of intensity and duration are negatively dependent (Zhang and Singh, 2012). The combined cumulative distribution function (CDF) in the copula similarly indicates that the two factors are negatively reliant on each other (Fig. 6).

The temporal probability of landslides (Table 2) over N years was calculated using joint conditional exceedance probability values. The conditional probability values show that, as the duration increases, the probability value decreases.

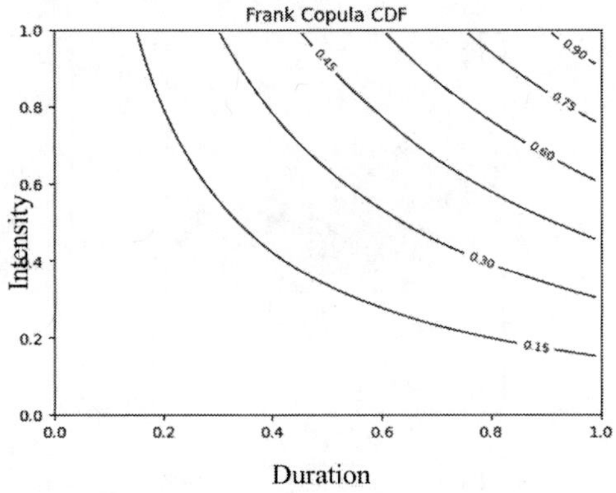

Figure 6. Contour plot of joint CDF of Mananthavady region.

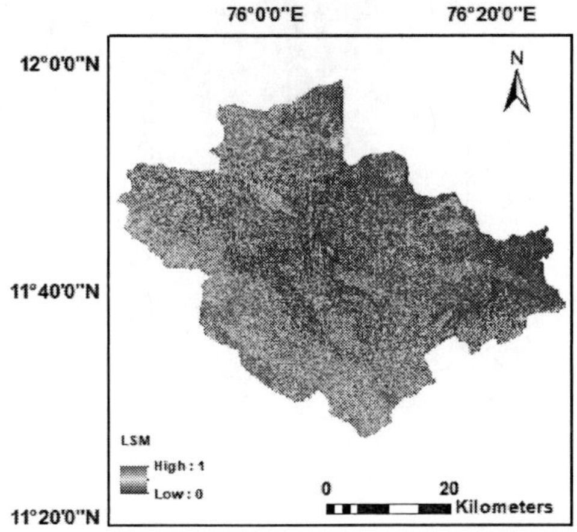

Figure 7. Landslide susceptibility map of Wayanad.

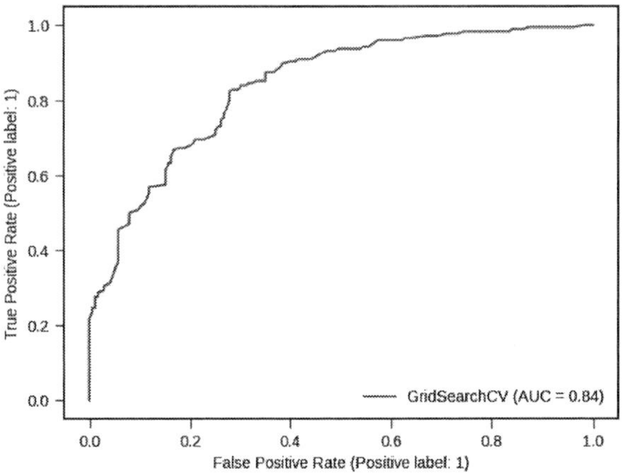

Figure 8. Area under the curve (AUC) for LSM.

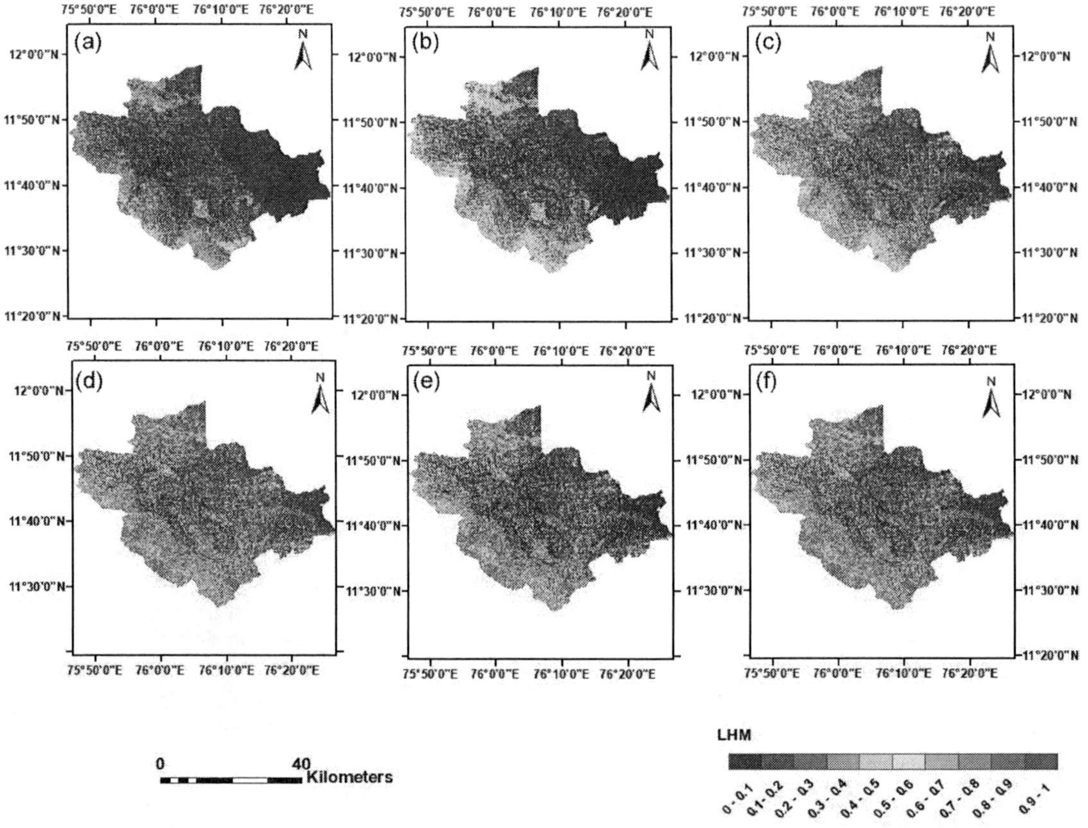

Figure 9. Landslide hazard maps for different periods of time: **(a)** 1 year, **(b)** 2 years, **(c)** 5 years, **(d)** 10 years, **(e)** 20 years, and **(f)** 50 years.

Table 2. Temporal probability of landslides over *N* years.

Location	Time period (years)					
	1	2	5	10	20	50
Mananthavady	0.49	0.74	0.97	1.00	1.00	1.00
Vythiri	0.54	0.79	0.98	1.00	1.00	1.00
Ambalavayal	0.58	0.83	0.99	1.00	1.00	1.00
Kuppady	0.18	0.32	0.62	0.86	0.98	1.00

Thus, the maximum exceedance value was observed for a 3 d period in the Mananthavady, Vythiri, Ambalavayal, and Kuppady regions. Equation (5) was then used to calculate the temporal probability for each region of Wayanad over various time periods.

4.3 Landslide susceptibility mapping

The map indicating the likelihood of landslides occurring (Fig. 7) was made by taking into account various factors that contribute to landslides. To confirm the accuracy of this map, the ROC curve was used. The LSM created using the random forest had an AUC value of 0.84, indicating a more accurate prediction, as shown in Fig. 8.

The feature selection shows that slope has the greater influence on landslide susceptibility, followed by SPI, TWI, and NDVI. The feature relevance of SPI and TWI demonstrates the indirect influence of rainfall on these parameters. The SPI shows the erosion power of water, whereas the TWI reflects soil saturation (Riihimäki et al., 2021; Mandal et al., 2021). The NDVI has an impact on the stream power index as well.

4.4 Landslide hazard mapping

The landslide hazard maps were prepared by integrating the temporal probability of landslides with the landslide susceptibility. The exceedance probability of each region was obtained as the temporal probability using the copula function, and then it was multiplied by the landslide susceptibility map to obtain the LHM (Fig. 9), showing the landslide hazard proneness calculated for time periods of 1, 2, 5, 10, 20, and 50 years. According to the hazard assessment, during a period of 1 to 2 years, the landslide hazard proneness was in the low and medium range of 70 %–80 % and 10 %–20 %, respectively. Over the period of 5 to 10 years, 30 %–40 % of the area is at medium hazard and 50 %–60 % is at moderate hazard. The area that is at high hazard proneness rises to 20 %–30 % after 10 years. The validity of the hazard map is proven when compared with the landslide occurrence on 3 July 2022 in the Vythiri region of Wayanad (https://www.manoramaonline.com/, last access: 29 June 2023). The results show that the landslide occurred in the area that was identified to be more hazard-prone, as predicted by the map. This confirms the importance of using such maps in preventing natural disasters and protecting lives and property.

5 Conclusions

The research discusses the use of a joint-probability approach to determine the temporal probability of landslides based on the intensity and duration of rainfall, the triggering factors. To determine the temporal probability, an empirical rainfall threshold was computed based on previous landslide occurrences. The research found that different regions of Wayanad such as Mananthavady, Vythiri, Ambalavayal, and Kuppady had varying rainfall intensities that could trigger landslides, including 19.08, 86.01, 18.24, and 25.47 $\mathrm{mm\,d^{-1}}$, respectively. The copula functions were used to obtain the temporal probability of landslides as the joint conditional exceedance probability of rainfall variables such as intensity and duration. The exceedance probability values obtained showed that, as the duration increased, the probability value decreased. Landslide hazard maps for different periods were prepared by integrating the landslide susceptibility and the temporal probability of landslides. The variation in the hazard proneness of different regions of Wayanad demonstrated the spatial relevance of multi-site rainfall information. The study concluded that the copula approach could analyze the combined effect of variables, thus reducing the impacts of natural hazards such as landslides.

Code and data availability. The code cannot be made publicly available at this time because the current work is still in progress. However, the authors can make the codes available with a genuine and reasonable request. All of the original data used in this study are freely available from their respective sources.

Author contributions. The work was conceptualized by SDS and AS. SDS retrieved and analyzed the data, as well as carried out the implementation. The draft was prepared by SDS. The paper was revised by AS. The research was overseen by AS.

Competing interests. The contact author has declared that neither of the authors has any competing interests.

Disclaimer. Publisher's note: Copernicus Publications remains neutral with regard to jurisdictional claims made in the text, published maps, institutional affiliations, or any other geographical representation in this paper. While Copernicus Publications makes every effort to include appropriate place names, the final responsibility lies with the authors.

Special issue statement. This article is part of the special issue "Mountain Hydrology and Cryosphere". It is a result of the International Conference on Mountain Hydrology and Cryosphere, Kathmandu and Dhulikhel, Nepal, 9–10 November 2023.

Acknowledgements. Shamla Dilama Shamsudeen acknowledges the Department of Science and Technology for recommending them for the WISE-KIRAN PhD Fellowship for sanctioning the project entitled "Regional hazard mapping of Western Ghats of Kerala, India accounting the temporal probability of rainfall" (grant no. DST/WISE-PhD/EA/2023/1, TPN-95768).

Financial support. This research has been supported by the Department of Science and Technology, Ministry of Science and Technology, India (grant no. DST/WISE-PhD/EA/2023/1, TPN-95768).

References

Abraham, M. T., Satyam, N., Rosi, A., Pradhan, B., and Segoni, S.: The Selection of Rain Gauges and Rainfall Parameters in Estimating Intensity-Duration Thresholds for Landslide Occurrence: Case Study from Wayanad (India), Water, 12, 1000, https://doi.org/10.3390/w12041000, 2020.

Achu, A. L., Joseph, S., Aju, C. D., and Mathai, J.: Preliminary analysis of a catastrophic landslide event on 6 August 020 at Pettimudi, Kerala State, India, Landslides, 18, 1459–1463, https://doi.org/10.1007/s10346-020-01598-x, 2021.

Ajin, R. S., Nandakumar, D., Rajaneesh, A., Oommen, T., Ali, Y. P., and Sajinkumar, K. S.: The tale of three landslides in the Western Ghats, India: lessons to be learnt, Geoenviron. Disast., 9, 16, https://doi.org/10.1186/s40677-022-00218-1, 2022.

Berti, M., Bernard, M., Gregoretti, C., and Simoni, A.: Physical Interpretation of Rainfall Thresholds for Runoff-Generated Debris Flows, J. Geophys. Res.-Earth, 125, e2019JF005513, https://doi.org/10.1029/2019jf005513, 2020.

Bezak, N., Šraj, M., and Mikoš, M.: Copula-based IDF curves and empirical rainfall thresholds for flash floods and rainfall-induced landslides, J. Hydrol., 541, 272–284, https://doi.org/10.1016/j.jhydrol.2016.02.058, 2016.

Crozier, M. J.: Deciphering the effect of climate change on landslide activity: A review, Geomorphology, 124, 260–267, https://doi.org/10.1016/j.geomorph.2010.04.009, 2010.

Haile, M. M.: Bivariate Flood Frequency Analysis: A case study of Rib River, Upper Blue Nile Basin, Ethiopia, ARRUS J. Eng. Technol., 2, 46–64, https://doi.org/10.35877/jetech489, 2022.

Hao, L., Rajaneesh A., van Westen, C., Sajinkumar K. S., Martha, T. R., Jaiswal, P., and McAdoo, B. G.: Constructing a complete landslide inventory dataset for the 2018 monsoon disaster in Kerala, India, for land use change analysis, Earth Syst. Sci. Data, 12, 2899–2918, https://doi.org/10.5194/essd-12-2899-2020, 2020.

Kim, H., Lee, J.-H., Park, H.-J., and Heo, J.-H.: Assessment of temporal probability for rainfall-induced landslides based on nonstationary extreme value analysis, Eng. Geol., 294, 106372, https://doi.org/10.1016/j.enggeo.2021.106372, 2020.

Lee, J. H., Kim, H., Park, H.-J., and Heo, J.-H.: Temporal prediction modeling for rainfall-induced shallow landslide hazards using extreme value distribution, Landslides, 18, 321–338, https://doi.org/10.1007/s10346-020-01502-7, 2020.

Mandal, K., Saha, S., and Mandal, S.: Applying deep learning and benchmark machine learning algorithms for landslide susceptibility modelling in Rorachu river basin of Sikkim Himalaya, India, Geosci. Front., 12, 101203, https://doi.org/10.1016/j.gsf.2021.101203, 2021.

Pabaghi, Z., Bazrafshan, O., Zamani, H., Shekari, M., and Singh, V. P.: Bivariate Analysis of Extreme Precipitation Using Copula Functions in Arid and Semi-Arid Regions, Atmosphere, 14, 275, https://doi.org/10.3390/atmos14020275, 2023.

Riihimäki, H., Kemppinen, J., Kopecký, M., and Luoto, M.: Topographic Wetness Index as a Proxy for Soil Moisture: The Importance of Flow-Routing Algorithm and Grid Resolution, Water Resour. Res., 57, https://doi.org/10.1029/2021wr029871, 2021.

Segoni, S., Piciullo, L., and Gariano, S. L.: A review of the recent literature on rainfall thresholds for landslide occurrence, Landslides, 15, 1483–1501, https://doi.org/10.1007/s10346-018-0966-4, 2018.

Shiau, J. T.: Fitting Drought Duration and Severity with Two-Dimensional Copulas, Water Resour. Manage., 20, 795–815, https://doi.org/10.1007/s11269-005-9008-9, 2006.

Zhang, L. and Singh, V. P.: Bivariate Rainfall and Runoff Analysis Using Entropy and Copula Theories, Entropy, 14, 1784–1812, https://doi.org/10.3390/e14091784, 2012.

Proc. IAHS, 387, 87–93, 2024
https://doi.org/10.5194/piahs-387-87-2024

Open Access PIAHS

Review on assessing climate-change-induced risks to run-of-river hydropower infrastructure in Nepal

Shraddha Kadel, Santosh Chaudhary, and Shyam Sundar Khadka

Department of Civil Engineering, Kathmandu University, 45210 Dhulikhel, Nepal

Correspondence: Shyam Sundar Khadka (sskhadka@ku.edu.np)

Published: 18 November 2024

Abstract. Nepal annually invests millions of dollars in hydropower development, with a substantial portion of these funds allocated to civil infrastructure. Climate change, which impacts geological and hydrological conditions, poses a threat to civil components throughout hydropower project construction and operation. This causes budget overruns, project delays, and adverse consequences for society and the environment, ultimately resulting in multimillion-dollar losses. The identification of risk factors and their underlying causes, stemming from the impacts of climate change, constitutes a fundamental aspect of this study. This critical analysis primarily draws upon extensive literature reviews to pinpoint these factors (risk factors arising from climate change), emphasizing their economic and human impacts. This paper highlights the risks to run-of-river hydropower infrastructures in Nepal. The findings of the study can be used to develop and implement adaptation strategies to mitigate the risks posed by climate change to Nepal's vital hydropower sector.

1 Introduction

Nepal's perennial rivers hold tremendous potential for hydropower development, offering a sustainable and environmentally friendly source of energy. Presently, hydropower plays a pivotal role in meeting approximately 96.2 % of Nepal's electricity needs (Basnet, 2022), and more than 90 % of hydropower is run-of-river type (Bhatt, 2017). Nepal took a significant step in the 1990s when frameworks, guidelines, regulatory bodies and acts were introduced (Nepal and Jamasb, 2012). This sector has attracted private investments exceeding NPR 55 billion (Lamsal, 2012). However, Nepal's geographical location within an earthquake-prone region and its proximity to the industrially advanced nations of India and China pose distinctive challenges. Nepal contributes a mere 0.1 % of global carbon emissions (World Bank, 2022), but it is in the fourth position with respect to vulnerability to the impacts of climate change (Joshi et al., 2019). This means that Nepalese regions covered in glacier and ice, encompassing 113 000 km^2 (Bhatt, 2017), are prone to glacial lake outburst floods (GLOFs), landslides and flash floods, especially in areas susceptible to seismic activity. In June of 2023, according to the Independent Power Producers' Association, Nepal (IPPAN), a total of 30 hydropower projects, with a combined capacity of 463 MW, were damaged. To protect the nation's investments in hydropower and ensure the safety of lives and property, it is important to establish a resilient foundation for hydropower development, considering the potential impacts of climate change; to avert future damage; and to effectively manage the associated risks (Wagle, 2022). As per the Nepal Electricity Authority (NEA) fiscal report of 2022/2023, there is a total hydropower project capacity of 2684 MW, with the commissioning of new projects equivalent to the capacity of 491 MW (NEA, 2023).

2 Literature review

2.1 Climate scenario, climate change and topography of nepal

Nepal's geography is distinctly partitioned into five major tectonic zones: the Terai, the Siwaliks, the Lesser Himalayas, the Higher Himalayas, and the Tibetan–Tethys zones (Upreti, 2001). This gives rise to five distinct climatic zones throughout the country, as illustrated in Fig. 1.

Based on climate data analysis from 1971 to 2014 (DHM, 2017), Nepal's annual maximum temperature trend is significantly positive (0.056 $^{\circ}C\,yr^{-1}$). The country's annual mini-

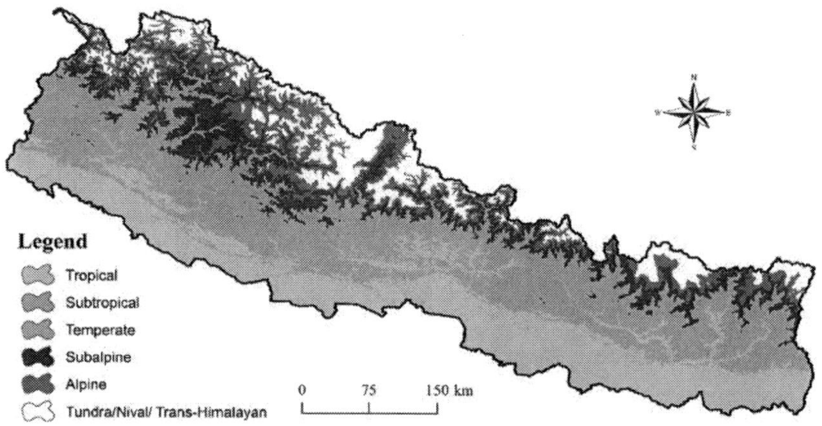

Figure 1. Climatic division of Nepal (source: Paudel et al., 2021).

mum temperature trend is also positive (0.002 °C yr^{-1}), although insignificant. No significant trend is observed in precipitation in any season. All five physiographic regions show a significant positive trend in all seasons, except for the Terai in winter and in the pre-monsoon season and the Siwaliks in winter. In high-mountain regions and the Higher Himalayas, the highest positive trend is observed in winter season, whereas the highest positive trend is observed in the monsoon season in the Terai, Siwaliks and middle-mountain regions.

Nepal has witnessed alterations in temperature and mean precipitation. Between 1975 and 2005, temperature data indicated an annual increase of 0.06 °C (Upadhayaya and Baral, 2020), while mean rainfall exhibited a significant decrease at an average rate of 3.7 mm (−3.2 %) per month per decade. Mean annual temperatures are anticipated to rise between 1.3 and 3.8 °C by the 2060s and by between 1.8 and 5.8 °C by the 2090s, while annual precipitation is expected to decline by 10 %–20 % nationwide (Bhatt, 2017). Multiple general circulation models converge with respect to projecting ongoing warming trends, with anticipated mean temperature increases of 1.2 and 3 °C by 2050 and 2100, respectively (Agrawala, 2003).

Figures 2 and 3 show two graphs with data about Nepal's climate. Figure 2 depicts the yearly mean temperature from 1981 to 2065, demonstrating a consistent rise in temperature. For the same time period, Fig. 3 indicates an increasing trend in precipitation levels. Both graphs demonstrate how the environment in Nepal is being affected by climate change.

Around 17 % of the Great Himalayas region is covered by glaciers and ice, spanning an area of nearly 113 000 km^2, with 35 000 km^2 dedicated to glaciers (Bhatt, 2017). There are 22 districts with a high vulnerability to landslides, 12 districts prone to GLOFs and 9 districts susceptible to flooding. Major threats are seen in Bhotekoshi, Tama Koshi, Marsyangdi, Arun and Dudh Koshi (NAPA, 2010).

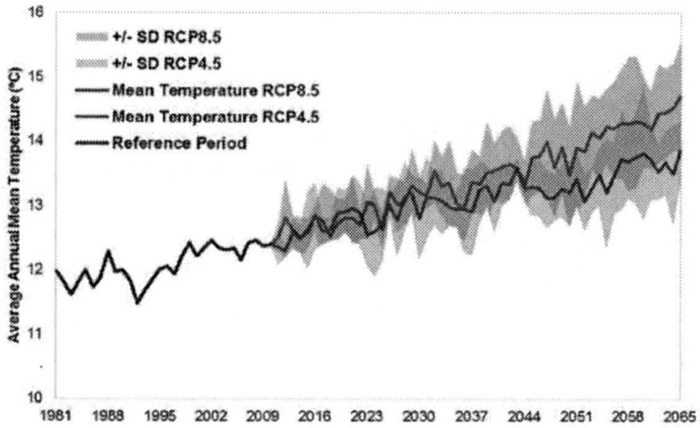

Figure 2. Annual mean temperature from 1981 to 2065; the chart estimates temperature changes from 0.9 to 1.1 °C (2016–2045) and from 1.3 to 1.8 °C (2036–2065) (source: Pradhananga and Nepal, 2020).

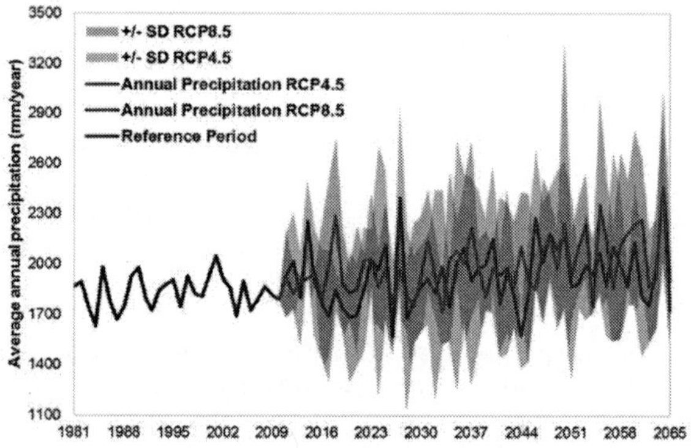

Figure 3. Annual precipitation from 1981 to 2065; the chart estimates precipitation changes from 2.1 % to 6.4 % (2016–2045) and from 7.9 % to 12.1 % (2036–2065) (source: Pradhananga and Nepal, 2020).

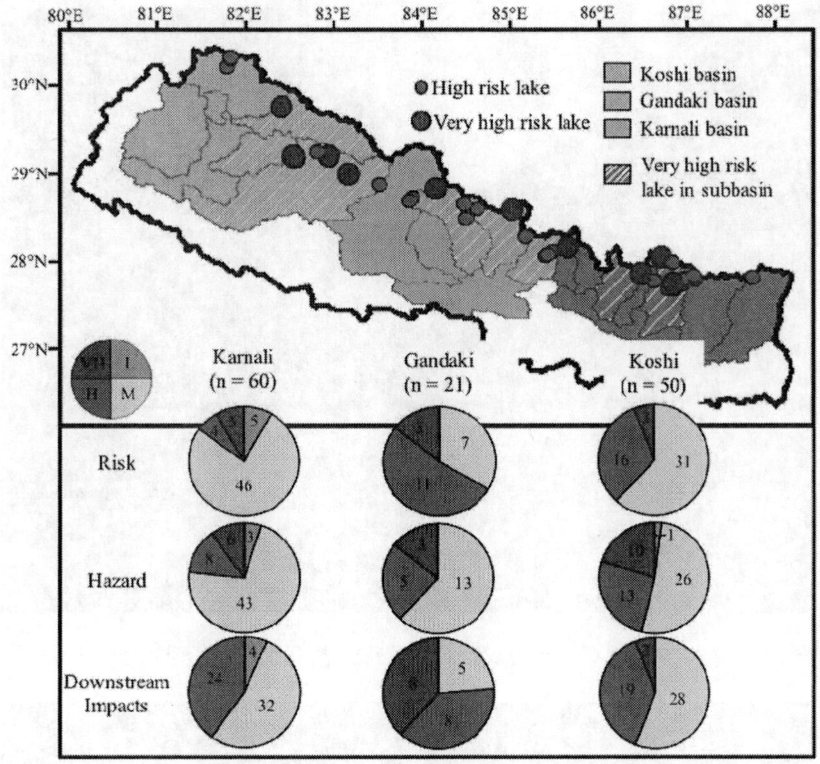

Figure 4. The distribution of risk, hazard and downstream impact classifications (pie charts) for each basin. Here, VH denotes very high risk, H denotes high risk, M denotes medium risk and L denotes less risk (source: Rounce et al., 2017).

A risk assessment map for lakes in the Karnali, Gandaki and Koshi basins is shown in Fig. 4. Lakes in high danger are denoted by red dots in each color-coded basin. The risk, hazard and downstream implications are shown in detail in the pie charts beneath the map. In general, this evaluation is essential for controlling the dangers connected to these lakes.

2.2 Impact of climate change on hydropower infrastructure

A rise in precipitation levels can trigger a cascade of challenges, including flash floods, erosion in catchment areas, landslides and slope failures, which pose a significant risk to critical components of hydropower infrastructure, such as reservoirs, intake structures and transmission lines. Intensifying heat can lead to the accelerated melting of glaciers, potentially culminating in GLOFs (Khanal, 2021).

Geological conditions and the risk of flooding were found to be the third and fourth most substantial challenges, respectively (Gurung, 2020). In the case of built–own–operate–transfer (BOOT) hydropower projects, geological risks emerged as the predominant concern in Nepal, often posing considerable disruptions during the construction phase (Tripathi and Shrestha, 2018). Research carried out in the Sapta Koshi River basin indicated that the flow pattern of the river has not varied much over the last 30 years, despite the effect of climate change (Khadka et al., 2018). There is a high sediment content in Nepal's rivers, coupled with topographical factors that increase the likelihood of rockfalls and landslides with localized particle properties potentially featuring fresh, sharp edges (Thapa et al., 2005).

Historically, Nepal has faced a series of devastating natural disasters. On 4 August 1985, there was a GLOF in Dig Tsho Lake, in the Langmoche Valley (Vuichard and Zimmermann, 1986), which led to the catastrophic destruction of the Namche hydroelectric project (HEP), damaging over USD 2–3 million and destroying 30 residential houses. Subsequently, on 27 August 2008, a massive flood swept away the 4 MW Khudi HEP which operated in 2007. This flood caused damage of NPR 8 440 147 (B. P. C., 2008).

On 2 August 2014, a catastrophic landslide in Jure created a dam blockage in a river, posing a significant threat to people living downstream. This disaster resulted in the loss of 10 % of the nation's hydropower capacity, amounting to 67 MW. It submerged a 5 MW power plant and severed connections with the 45 MW Bhotekoshi HEP and 10 MW

Figure 5. The Jure landslide damaged the road and headwork of the Sunkoshi hydropower project (source: Panthi, 2021).

Figure 6. Impacts of GLOFs on the Upper Bhotekoshi Koshi HEP reservoir (source: Bhatt, 2017).

Sunkoshi HEP. The landslide led to the destruction of over 400 houses, claimed the lives of 156 people and resulted in damage totaling more than USD 25 000 (van der Geest, 2018; Bhatt, 2017). Figure 5 illustrates the damage caused by the Jure landslide in the headworks of the Sunkoshi HEP.

The 7.8 (on the Richter scale) earthquake that struck Nepal on 25 April 2015 resulted in significant repercussions with respect to the country's hydropower infrastructure: more than 14 hydropower dams were damaged (Schneider, 2015). The 24 MW Trishuli 3A plant was hit by a landslide that killed four people and formed cracks in the reservoir crest of the Kulekhani HEP (Schneider, 2015). The 14 MW plant located downstream at Devighat was rendered nonoperational until the Trishuli 3A plant could be restarted (Fairley, 2015). A total of 14 of the 34 hydropower projects were damaged between 2015 and 2020. The 200 kW Malagad hydropower project was damaged in the floods of 2020, impacting 46 powered factories (UNDP, 2023).

On 6 July 2016, a landslide dam in Nyalam, Tibet, burst, leading to the Upper Bhotekoshi HEP, with a capacity of 36 MW, being swept away and resulting in infrastructure damage totaling USD 406.739 million (Khanal et al., 2015; Bhatt, 2017). Figure 6 depicts the damage incurred in the hydropower reservoir of Upper Bhotekoshi HEP.

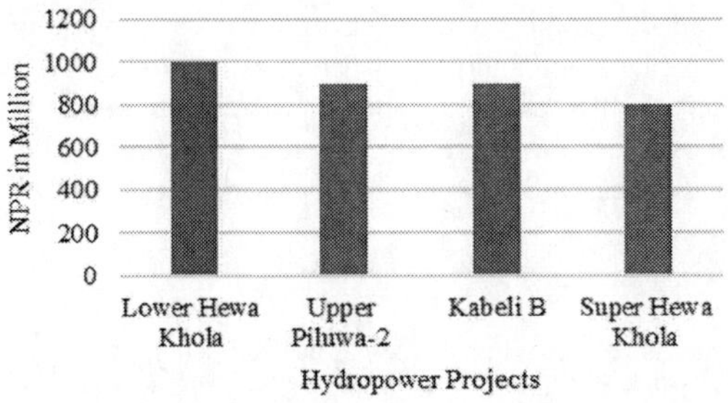

Figure 7. Bar graph showing the damage to major hydropower projects due to the flash flood which occurred in June 2023 (source: *The Annapurna Express* newspaper, 26 June 2023; the *República* newspaper, 24 June 2023).

Figure 8. Damage due to the flood on 16 and 17 June 2023 (source: the *República* newspaper, 24 June 2023).

Figure 9. Damage to the Hewa HEP due to the flood on 16 June 2018 (source: *The Annapurna Express* newspaper, 26 June 2023).

The frequency of floods and landslides has resulted in significant damage along roads, impeding the flow of construction supplies. As such, there have been notable setbacks in the development of multiple hydroelectric projects. The impact of floods has been severe, according to IPPAN, damaging 16 active projects and causing operational disruptions to 10 electricity-generating projects. It is believed that the damage resulting from landslides and floods brought on by continuous rains has amounted to approximately NPR 3 billion. The Super Madi Hydropower project in Madi, Kaski, is one of these, and it alone has lost more than NPR 1 billion, demonstrating the extent of the difficulty that the construction industry faces.

As reported by IPPAN, the floods that occurred on 16 and 17 June 2023 in the Taplejung, Panchthar, Sankhuwasabha and Bhojpur districts resulted in substantial damage to a total of 30 hydropower projects, with a combined capacity of 463 MW. Figure 7 illustrates some of the major affected hydropower projects.

From a total of 30, 13 operating projects with a cumulative capacity of 132 MW incurred losses of approximately NPR 6 billion, while 17 hydropower projects under-construction, with a total capacity of 327 MW, suffered losses exceeding NPR 2 billion. A total of 18 workers were swept away by the flood at the Super Hewa Khola project site. The hydropower plants faced a total damage bill of NPR 8 billion. Figure 8 illustrates the damage caused by the floods on 16 and 17 June. Figure 9 depicts the damage to the Hewa HEP due to the flood on 16 June 2023.

In April 2023, a landslide occurred at the construction site of the 900 MW Arun III hydropower project, creating the potential for a flash flood. To mitigate this risk, immediate action was taken to shut down the tunnel. Figure 10 shows the blockage at the Arun III Dam site.

The flood on 4 October 2023, resulting from a glacial lake outburst originating from South Lhonak in Sikkim, destroyed the 1200 MW Chungthang Dam. This incident serves as a stark reminder of the growing threat to Nepal's hydropower projects, including Tamor, Arun and Kabeli, currently under construction. The damage done to Chungthang Dam is shown in Fig. 11.

3 Research findings

Nepal's mean annual temperature has been rising by 0.06 °C annually, and the country's Himalayan area is experiencing a higher rate of climate change than the rest of the world (Upadhayaya and Baral, 2020). Projections suggest a temperature rise of 1.2–3 °C and a precipitation decrease of 10 %–20 % by 2050 (Bhatt, 2017). The impact of climate change is strikingly evident in Nepal's hydropower sector, which currently boasts an operational capacity of 2684 MW, with the commissioning of new projects equivalent to a capacity of 491 MW that are heavily dependent on run-of-river sources (NEA, 2023). The country has faced increased flooding, landslides and glacial outbursts. As reported by IPPAN, the floods that occurred on 16 and 17 June 2023 alone resulted in damage totaling a capacity of 463 MW to hydropower

Figure 10. A diversion channel blocked at the Arun III Dam site due to a landslide (source: *The Kathmandu Post* newspaper, 1 May 2023).

Figure 11. Damage to Chungthang Dam (source: the *Nepali Times* newspaper, 4 October 2023).

projects and financial loss of over NPR 8 billion. Escalating climate-induced volatility raises concerns about outbursts of glacial lakes, which cover 35 000 km^2 of the nation's land. This underscores the urgent need for an early-warning system, design parameter adaptations and increased investment to mitigate climate change effects on the hydropower sector.

4 Conclusion

In conclusion, research findings highlight Nepal's urgent climate change challenges, especially with respect to the hydropower sector. The increased temperatures and reduced precipitation, along with the anticipated changes in the country's environment, underscore the vulnerability of the nation's hydropower and reservoir infrastructure. Significant financial losses and human casualties have resulted from the recent catastrophic occurrences, including landslides, floods and glacial outbursts. The increasing effects emphasize how crucial it is to take quick action, such as modifying design specifications, putting an efficient early-warning system in place and increasing funding for climate resilience initiatives. It is critical to address these problems to protect Nepal's hydropower industry, promote sustainable development and lessen the wider socioeconomic effects of climate change.

Data availability. This study utilized data from journal articles, book sections, annual reports, conference proceedings, and newspapers. Data sources include the Nepal Electricity Authority (NEA, 2023, https://www.nea.org.np/annual_report), Department of Hydrology and Meteorology (DHM, 2017, https://www.climatenepal.org.np/resources/observed-climate-trend-analysis-nepal-1971-2014), National Adaptation Programme of Action (NAPA, 2010, https://adaptation-undp.org/resources/assessments-and-background-documents/nepal-national-adaptation-programme-action-napa) and various newspapers, all of which are cited in the references and figure captions.

Author contributions. The foundational concept was formulated by SSK, under whose guidance SK conducted data collection and review. SC and SK carried out the writing and editing processes.

Competing interests. The contact author has declared that none of the authors has any competing interests.

Special issue statement. This article is part of the special issue "Mountain Hydrology and Cryosphere". It is a result of the International Conference on Mountain Hydrology and Cryosphere, Kathmandu and Dhulikhel, Nepal, 9–10 November 2023.

Acknowledgements. We thank the Department of Civil Engineering for this opportunity and for funding for paper publication.

References

Agrawala, S.: Development and climate change in Nepal: Focus on Water Resources and Hydropower, Organisation for Economic Co-operation and Development, https://inis.iaea.org/search/search.aspx?orig_q=RN:36000404 (last access: 7 June 2024), 2003.

Basnet, S.: The role of and challenges and prospects for hydropower development in Nepal's energy sector, Nepal Econ. Forum, https://nepaleconomicforum.org/the-role-of-and-challenges-and-prospects-for-hydropower-development-in-nepals-energy-sector (last access: 18 July 2024), 2022.

Bhatt, R. P.: Hydropower Development in Nepal-Climate Change, Impacts and Implications, in: Renewable Hydropower Technologies, 3rd edn., edited by: Rajakaruna, M., IntechOpen, https://www.intechopen.com/chapters/53350 (last access: 18 July 2024), 2017.

B. P. C.: Annual Report 2008, Buddha Nagar, Kathmandu: Butwal Power Company Limited, https://www.bpc.com.np/investor-relations/annual-report (last access: 18 July 2024), 2008.

DHM: Observed Climate Trend Analysis in the Districts and Physiographic Regions of Nepal (1971–2014), Department of Hydrology and Meteorology, Kathmandu, https://www.climatenepal.org.np/resources/observed-climate-trend-analysis-nepal-1971-2014 (last access: 18 July 2024), 2017.

Fairley, P.: Nepal Hydropower Battered but not Beaten by the quake, IEEE Spectrum, https://spectrum.ieee.org/nepals-hydropowerbased-power-system-survived-its-m79-quake-(last access: 18 July 2024), 2015.

Gurung, S.: Risk Analysis for the Sustainable Hydropower Development in Nepal, Int. J. Adv. Eng. Sci. Appl., 1, 1–10, https://doi.org/10.47346/ijaesa.v1i2.39, 2020.

Joshi, A., Farquhar, S., Assareh, N., Dahlet, L., and Landahl, E.: Climate change in Lamjung District, Nepal: meteorological evidence, community perceptions, and responses, Environ. Res. Commun., 1, 025005, https://doi.org/10.1088/2515-7620/ab1762, 2019.

Khadka, S. S., Bhattarai, U., and Maskey, R. K.: Standardization of Civil Design Framework for Small Hydropower Projects: Assessing Existing Design Procedures and Considering Impacts of Climate Change, in: Proceedings of the 2018 International Conference on Developments in Renewable Energy Technology (ICDRET'2018), https://www.researchgate.net/publication/344167720_Standardization_of_Civil_Design_Framework_for_Small_Hydropower_Projects_Assessing_Existing_Design_Procedures_and_Considering_Impacts_of_Climate_Change (last access: 18 July 2024), 2018.

Khanal, N., Hu, J.-M. and Mool, P.: Glacial Lake Outburst Flood Risk in the Poiqu/Bhote Koshi/Sun Koshi River Basin in the Central Himalayas, Mt. Res. Dev., 35, 351–364, 2015.

Khanal, S.: Climate Risk and Adaptation Assessment for Hydropower Project in Nepal, Future Water, https://www.futurewater.eu/projects/climate-risk-and-adaptation-assessment-for-hydropower-project-in-nepal (last access: 18 July 2024), 2021.

Lamsal, M.: Private Sector Investment In Hydropower Reaches Rs 55 Billion, Kathmandu, https://www.newbusinessage.com/articles/view/1541 (last access: 18 July 2024), 2012.

NAPA: National Adaptation Programme of Action, Nepal Climate Change and Development Portal, https://adaptation-undp.org/resources/assessments-and-background-documents/nepal-national-adaptation-programme-action-napa (last access: 12 May 2024), 2010.

NEA: Nepal Electricity Authority a year in a review-fiscal year 2022/2023, Durbar Marg, Kathmandu: Nepal Electricity Authority, https://www.nea.org.np/annual_report (last access: 18 July 2024), 2023.

Nepal, R. and Jamasb, T.: Reforming small electricity systems under political instability: The case of Nepal, Energy Policy, 40, 242–251, 2012.

Panthi, K. K.: Assessment on the 2014 Jure Landslide in Nepal – a disaster, Mech. Rock Eng. Theory Pract., 833, 012179, https://doi.org/10.1088/1755-1315/833/1/012179, 2021.

Paudel, B., Pandaya, D., and Dhakal, K.: Climate, The Soils of Nepal, 15–27, https://www.springerprofessional.de/en/climate/19759384 (last access: 18 July 2024), 2021.

Pradhananga, S. and Nepal, S.: Climate change scenarios, ICIMOD, Kathmandu: Regional climate change projections: Climate change, https://www.icimod.org/wp-content/uploads/2020/11/Day2_Climate_Change_NAP_CORDEX_data_analysis.pdf (last access: 18 July 2024), 2020.

Rounce, D. R., McKinney, D. C. and Watson, C. S.: Identification of Hazard and Risk for Glacial Lakes in the Nepal Himalaya Using Satellite Imagery from 2000–2015, Remote Sens., 9, 654, https://doi.org/10.3390/rs9070654, 2017.

Schneider, K.: Nepal Earthquake Damages At Least 14 Hydropower Dams, Circle Blue, WaterNews, https://www.circleofblue.org/2015/world/nepal-earthquake-damages-at-least-14-hydropower-dams/ (last access: 18 July 2024), 2015.

Thapa, B., Shrestha, R., Dhakal, P. and Shrestha, B. S.: Problems of Nepalese hydropower projects due to suspended sediments, Aquat. Ecosyst. Health Manage., 8, 373–379, 2005.

Tripathi, P. and Shrestha, S. K.: Risk Assessment of Boot Hydropower Projects in Nepal Using Fuzzy Logic Approach, J. Adv. Coll. Eng. Manage., 4, 9–16, 2018.

UNDP: How energy is transforming communities in Bajura, UNDP Nepal, https://www.undp.org/nepal/stories/how-energy-transforming-communities-bajura (last access: 18 July 2024), 2023.

Upadhayaya, R. P. and Baral, M. P.: Trends of Climate Change in Some Selected Districts of Western Nepal, Janapriya J. Interdiscip. Stud., 9, 148–158, 2020.

Upreti, B. N.: The physiographic and geology of Nepal and their bearing on the landslide problem, in: Landslide Hazard Mitigation in the Hindu Kush-Himalaya, edited by: Li, T. C., Chalise, S. R., and Upreti, B. N., Kathmandu, International Centre for Integrated Mountain Development, 31–49, 2001.

van der Geest, K.: Landslide Loss and Damage in Sindhupalchok District, Nepal: Comparing Income Groups with Implications for Compensation and Relief, Int. J. Disaster Risk Sci., 9, 157–166, 2018.

Vuichard, D. and Zimmermann, M.: The Langmoche Flash-Flood, Khumbu Himal, Nepal, Mt. Res. Dev., 6, 90–94, 1986.

Wagle, N.: Breaking the bottlenecks: Managing multi-hazard risks in the hydropower sector, Lalitpur: ICIMOD, https://www.icimod.org (last access: 18 July 2024), 2022.

World Bank: Integrating Climate Change into Nepal's Development Strategy Key to Build Resilience, Says New World Bank Group Report, https://www.worldbank.org/en/news/press-release/2022/09/15/ (last access: 18 July 2024), 2022.

Proc. IAHS, 387, 95–101, 2024
https://doi.org/10.5194/piahs-387-95-2024

Open Access
PIAHS

Assessment of permafrost-related hazards in China: based on Chinese literature

Silian Pan, Prashant Baral, and Miriam Jackson

International Centre for Integrated Mountain Development (ICIMOD), Lalitpur, Nepal

Correspondence: Silian Pan (silianpan@outlook.com)

Published: 18 November 2024

Abstract. High Mountain Asia (HMA) is undergoing unprecedented warming, affecting the cryosphere – including permafrost (frozen ground) – and leading to various hazards. However, understanding the prevalence, distribution, and dynamics of these hazards and how they respond to a changing climate is challenging. Permafrost is extensive in HMA, and China makes up a significant portion of this. The permafrost area in China is about 1.6×10^6 km^2, 66 % of which is on the Qinghai–Tibet Plateau. However, most of the scientific literature concerning permafrost in China is published in Chinese and, hence, remains largely unnoticed by the non-Chinese-speaking scientific communities. In this article, we used a systematic review to evaluate the Chinese scientific literature on permafrost-related hazards and found that the studied areas are concentrated in certain areas, especially on the Qinghai–Tibet Engineering Corridor (QTEC). The increasing amount of literature on permafrost hazards reflects the increased impact of climate warming on infrastructure built on permafrost. Not only is permafrost affecting infrastructure; these anthropogenic disturbances themselves also have amplified the occurrence of hazards around settlements and infrastructure. The literature shows the strong relationship between latitude and elevation with permafrost thickness. The permafrost classification system and nomenclature used by Chinese scientists is different to that used elsewhere, which is a potential source of confusion and deserves attention.

1 Introduction

High Mountain Asia (HMA) is undergoing unprecedented warming (Yao et al., 2019; Hock et al., 2019), causing considerable changes to the cryosphere (Bolch et al., 2019), including to the permafrost in high mountains and on mountain plateaus (Li et al., 2022). China constitutes a large portion of HMA, including several high mountain ranges, with a significant distribution of glaciers and permafrost; transboundary river basin systems; and the highest plateau in the world, the Qinghai–Tibetan Plateau (QTP). In China, about 1.6×10^6 km^2 is considered to be permafrost; about 66 % of this in the QTP, 15 % is in northeastern China, and 19 % is in the mountains of the western and central parts of the country (Ran et al., 2012).

Permafrost is commonly defined as frozen ground at or below 0 °C for at least 2 consecutive years (Dobinski, 2011). It is a thermal phenomenon (Gruber Stephan and Haeberli, 2009) and is often identified using ground surface temperature measurements or visible indicators such as rock glaciers or other slow-moving bodies of rock and ice that exist in permafrost environments. Globally, permafrost areas are responding to higher air temperatures; with higher ground temperatures and the thickening of the active layer or the soil portion above permafrost that undergoes seasonal freezing and thawing (Dobiński, 2020), permafrost areas are thawing, warming, degrading, and disappearing, consequently increasing the risks of rockfalls, debris flows, thaw slumps, and permafrost-dammed lake outbursts (Ding et al., 2021; Biskaborn et al., 2019). Permafrost warming and degradation influence not only cryosphere-related landforms but also fluvial landforms and processes by controlling the surface runoff, by eroding sediments, and through the development of thermo-erosional gullies (Tananaev and Lotsari, 2022). The degradation of permafrost holds particular significance in mountainous regions due to its profound impact on slope stability. Consequently, this process modifies the geophysical attributes of mountain slopes, diminishing cohesive forces and elevating hydraulic conductivity (Patton et al., 2019). Water deposits affecting hydrothermal exchanges, fluvio-thermal erosion due to surface runoff, and swelling and subsidence of frozen grounds due to frost heave and thaw settlements strongly affect infrastructure built on thawing-permafrost grounds.

Understanding the probability and dynamics of potential risks due to permafrost thaw are vital for hazard reduction and safeguarding critical infrastructure (Harris et al., 2009). The QTP holds significant importance in permafrost research, primarily because it houses critical infrastructures like the Qinghai–Tibet railway and Qinghai–Tibet highway. Within China, numerous specialized research programs are actively studying the interplay between changing permafrost conditions and their impact on infrastructure, as well as the reciprocal effects of infrastructure on permafrost (Wu et al., 2010; Cheng and Jin, 2013; Ran et al., 2022). There is extensive Chinese literature and a comprehensive database, such as the China National Knowledge Infrastructure, containing an extensive compilation of permafrost-related publications and summaries. This wealth of knowledge is often overlooked and excluded in systematic assessments and reviews of permafrost changes in High Mountain Asia (Baral et al., 2023). Our systematic review tries to advance the scientific knowledge related to climate change impacts on permafrost dynamics and the occurrence of hazards related to thawing permafrost discussed in Chinese literature by making it more accessible to the English-speaking scientific community. This study attempts to answer the following questions to reveal to a wider audience the information stored in the Chinese literature database:

- According to Chinese scientific literature, what are the major hazards associated with changing permafrost in China and how are they distributed across Chinese territories within HMA?
- What are the major particularities in the Chinese literature that are different from the English literature?
- Are specific trends related to the occurrence of hazards visible?
- What future concerns are related to infrastructure construction and maintenance in the context of global warming and changes in permafrost in China?

Our results are totally based on the outcome of the systematic review of the Chinese literature, while the Introduction and Discussion sections are largely based on the information about permafrost-related hazards in HMA, mainly in China, in the English literature. In some cases, we mention the Hindu Kush Himalaya (HKH), referring to the region ranging from 15.95 to 39.31° N and 60.85 to 105.04° E, which includes the high-elevation regions in China, Bhutan, India, Pakistan, Nepal, Afghanistan, Bangladesh, and Myanmar (Bajracharya and Shrestha, 2011).

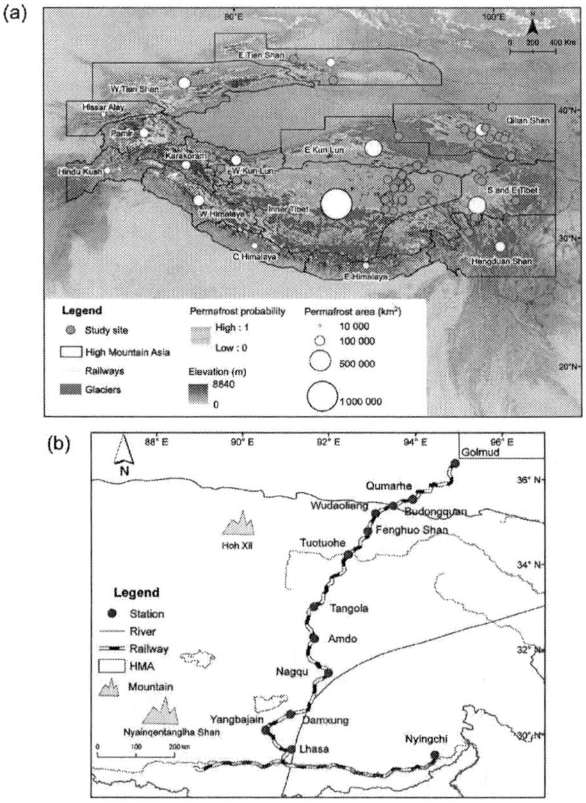

Figure 1. Study sites in High Mountain Asia **(a)** and Qinghai–Tibet Engineering Corridor **(b)**.

2 Methods

2.1 Study area

To understand permafrost and its related hazards, our research is centered in the ecologically significant region of High Mountain Asia. This vast area encompasses some of the world's highest and most imposing mountain ranges, including the Himalayas, Karakoram, Hindu Kush, and Pamir. HMA is characterized by its extreme elevations, low temperatures, and rugged terrain, making it a prime location for the study of permafrost dynamics and associated risks.

Most of the study sites were located on the Qinghai–Tibet Plateau, an integral part of the HMA region, often referred to as the "Third Pole" due to its immense ice reserves. The QTP is the highest and largest plateau on Earth and is situated in western China and parts of Tibet. Its unique and remote landscape presents an ideal setting for the study of permafrost due to its extensive permafrost coverage and sensitivity to climate change. The Qinghai–Tibet Engineering Corridor (QTEC) encompasses a network of critical infrastructure projects, including highways, railways, and energy

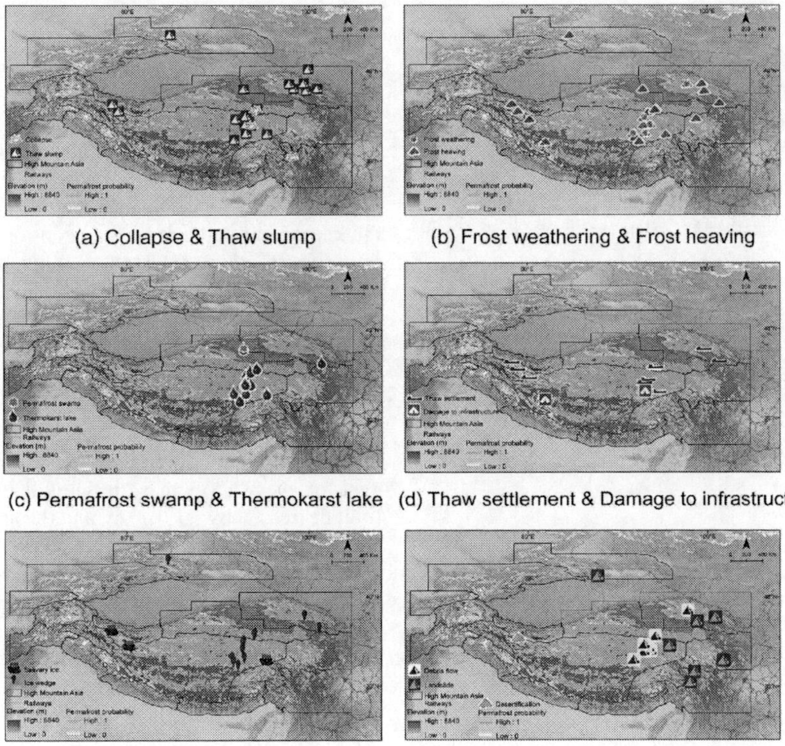

(a) Collapse & Thaw slump (b) Frost weathering & Frost heaving
(c) Permafrost swamp & Thermokarst lake (d) Thaw settlement & Damage to infrastructure
(e) Salivary ice & Ice wedge (f) Debris flow & Landslide & Desertification

Figure 2. Distribution by hazard type.

developments, spanning the vast and challenging terrain of the QTP. This corridor presents a unique opportunity to study the interaction between human activities and the delicate permafrost environment.

2.2 Systematic literature review method

This study built on the principles of the systematic review approach to assess the Chinese-language literature on permafrost hazards in High Mountain Asia and included the influence of structures on permafrost (James et al., 2016; Haddaway and Macura, 2018). In sustaining the replicability and transparency of systematic reviews, we adopted the RepOrting standards for Systematic Evidence Syntheses (ROSES) framework for ensuring evidence synthesis; this framework presents a protocol for searching, screening, and synthesis (Fig. S1 in Pan et al., 2024, https://doi.org/10.5281/zenodo.12157179). The following relevant indicators were included in the compilation: site information (coordinate, temperature), permafrost (thickness, active-layer thickness), hazard category, and infrastructure (type, area, length).

3 Results

3.1 Distribution of study areas

The place with the most hazard records was the QTEC (Fig. 2). Obviously, Chinese literature paid more attention to this area. A total of 10 distinctive hazards took place in this engineering corridor, including collapse, thaw slumps, frost weathering, frost heaving, thermokarst lake, thaw settlement, damage to infrastructure, ice wedge, debris flow, and landslides (e.g., Luo et al., 2014; Ma et al., 2012). Qilian Shan mainly suffered from thaw slumps, frost weathering, frost heaving, thaw settlements, ice wedges, debris flows, and landslides (e.g., Mu, 2021; Pang et al., 2019; Shang, 2018). Western Kunlun Shan experienced thaw slumps, frost heaving, thaw settlements, salivary ice, and desertification (e.g., Ma et al., 2004; Zhu, 2017). Tien Shan, which is outside the QTP but located in HMA, was found to be affected by thaw slump, frost heaving, ice wedge, and landslide (e.g., Ci et al., 2006). Besides this, thermokarst lakes, ice wedges, debris flows, and landslides were exclusively recorded in the east of the QTP.

All field sites were clustered into four zones around the QTEC, Qilian Shan, western Kunlun Shan, and Tien Shan.

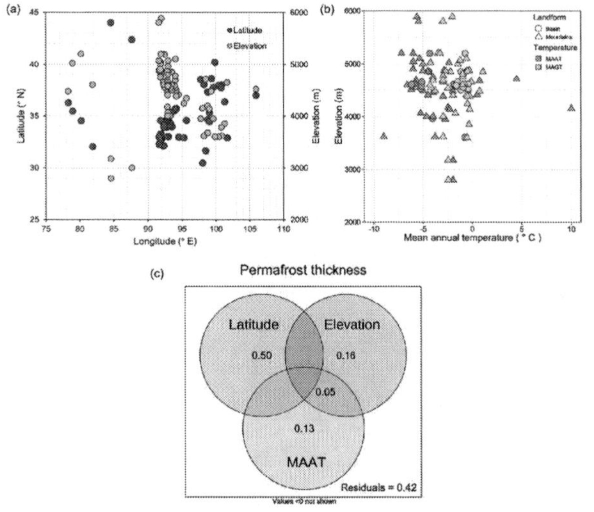

Figure 3. Coordinates and elevations given in publications **(a)**. Elevation-dependent mean annual air and ground temperature changes over HMA **(b)**. Variation partition analysis (VPA) revealing the impacts of latitude, elevation, and mean annual air temperature on permafrost thickness **(c)**. The values given in the non-overlapping parts of each circle give the weight of each factor. The value given in the overlapping circles gives the explanation of the variation between two factors.

The elevation gradient was between 2800 to 5886 m. The latitude and longitude ranges of these sites were 30.5 to 44.0° N and 78.3 to 105.9° E (Fig. 3a). Field-based research sites in the mountains are distributed within the wide-ranging elevation from 2800 m to almost 6000 m a.s.l., while research sites in the river basins spread from around 3000 to 5200 m a.s.l. (Fig. 3b). Consequently, mean annual air temperature (MAAT hereafter) and mean annual ground temperature (MAGT hereafter) values recorded at the study sites in the mountains have a wider range of values compared to those of the study sites in the river basins. The distribution of MAAT in the river basins varies from −6.24 to −0.5 °C, while in the mountains, MAAT ranges from −9 to −4.3 °C. MAGT has a very narrow range in the basins (−1.75 to −0.5 °C) compared to in the mountains (−4.25 to 1.1 °C).

Active-layer thickness varies from 1 to 6 m for MAGT ranging from −5 to slightly more than 1 °C; study sites record MAAT between −8 and 1 °C (Fig. S3 in Pan et al., 2024). The thickness of permafrost reported for multiple field sites ranges between 0 to 200 m. In locations where MAGT ranges from −2 to 0 °C, permafrost thickness is usually below 50 m. Permafrost thickness is mostly reported for sites with MAAT from −6 to −3 °C.

3.2 Influence of different parameters

Coordinates, topography, and temperature all show a correlation with the variance in permafrost thickness (Wang et al., 1979; Liu, 2008; Yin et al., 2014). In particular, latitude, elevation, and mean annual air temperature all independently explained the variance in permafrost thickness. The best model for understanding the controls on permafrost thickness indicated that latitude controlled 50 % of the variation, and elevation explained 16 % independently of the 13 % explained by the MAAT. Elevation and MAAT shared a small amount of variance in terms of permafrost thickness (5 %), confirming the MAAT specificity of elevation (Fig. 3c). The permafrost on the Qinghai–Tibet Plateau was influenced by elevation and followed the latitudinal zonation pattern (Zhou et al., 2000). The law of decreasing the lower limit of permafrost with increasing latitude was quite evident. On average, for every 1° increase in latitude, the permafrost lower limit decreased by 80–100 m, and the annual average ground temperature decreased by 0.9–1 °C (Wang et al., 1979; Bian et al., 2011). There was also a solid vertical zonation pattern of increasing permafrost thickness with higher elevation. For approximately every 100 m increase in elevation, the annual average ground temperature decreased by 0.5–0.6 °C, and the permafrost thickness increased by 15–20 m (Bian et al., 2011).

4 Discussion

4.1 Permafrost in English and Chinese scientific literature

In China, permafrost has been systematically investigated since the 1950s (Zhang, 2005). Development of permafrost science in China has been closely associated with the infrastructure and economic development; therefore, there are dedicated institutions in China with decades of intensive research. Permafrost research in China is based on standard definitions and classifications (Tables S1, S2, and S3 in Pan et al., 2024) commonly accepted by Chinese institutions and researchers. These standard definitions and classifications do not always match with the contemporary definitions and classifications accepted by the English scientific community (Ran et al., 2012). For example, the International Permafrost Association defines the percentage of permafrost for continuous, discontinuous, sporadic, and isolated patches to be 90 %–100 %, 50 %–90 %, 10 %–50 %, and 0 %–10 %, respectively, while some of the Chinese literature uses continuous, discontinuous, and island permafrost with > 65 %, 30 %–65 %, and 5 %–30 %, respectively (Qinglai, 2007). Similarly, although the classification of permafrost based on ice content is well established in the English literature, the Chinese literature uses the classification based on the ice content criterion (Table S1 in Pan et al., 2024), termed the permafrost engineering classification. Additionally, there are several Chinese terms that are commonly used in Chinese scientific literature but that are not common or are not used at all in English scientific literature. Some examples of such are Chinese terms whose literary translations in English are "salivary ice", "extruded ice", and "ice mantle" (e.g., Liu et al., 2003; Liu, 2008; Ma et al., 2004).

4.2 Permafrost hazards

Thermokarst terrains, such as retrogressive thaw slumps, are increasing freeze–thaw erosions and modifying landscapes on the Qinghai–Tibet Plateau (QTP). Permafrost warming has extensively increased permafrost collapse events across the Qinghai–Tibetan Plateau, affecting pedological characteristics and soil microbial communities (Hu and Yang, 2023).

Permafrost degradation has some major implications for infrastructure: damages to embankments, failure of bridges, and longitudinal cracks along highways (Hjort et al., 2022). Large-area coverage of permafrost (about two-fifths of the total area) on the QTP includes large sections of highways and several thousands or kilometers of railway lines; notably, from 1991 to 2011, almost USD 1 billion were spent on the maintenance of QTP. The cost for mitigation of damages to ensure the smooth functioning of highways and railway networks in China could reach up to several billions of US dollars between 2051–2100 (Hjort et al., 2022).

Regular field-based monitoring of meteorological variables using automated weather stations, soil moisture and temperature measurements where infrastructure foundations are located, regular surveys of the ground subsidence, stability and support assessments, and effective implementation of early-warning systems are critical for infrastructure located in thawing-permafrost regions (Wang et al., 2019).

Changes in ground ice strongly determine risks for communities on the QTP. Simulations suggest that permafrost area on the QTP decreased by about 0.25×10^6 km^2, while ground temperatures increased by 0.25 °C per decade between 1981 and 2010 (Xu et al., 2017). Freeze–thaw hazards and thaw settlements due to ground ice changes will have impacts on existing and planned infrastructure and so will have subsequent economic implications. Elevation-dependent temperature response will considerably increase the active-layer thickness and decrease permafrost areas on the Tibetan Plateau (Wang et al., 2022), increasing the risks of geohazards.

A more detailed discussion about permafrost hazards, adaptation to these hazards, and future projections can be found in Pan et al. (2024) (Sup_Assessment of permafrost-related hazards in China based on Chinese literature.docx)

5 Conclusions

The volume of scientific literature is increasing; the QTEC is an obvious geographical preference for conducting measurements and assessments of permafrost-related hazards. Studies and measurements are concentrated around a few hotspot areas, while extensive areas in China remain unexamined.

Projections of hazards based on complex modeling approaches using projected socioeconomic global changes are limited. Expanding measurements and research will be needed to provide a comprehensive picture of the dynamics of changing permafrost and potential hazards.

Increasing warming is likely to invite climate extremes and increase the frequency of cascading and compounding events linked to permafrost thawing. Human populations and infrastructures exposed to permafrost hazards will increase as a consequence of increasing human activities in changing permafrost areas.

Finally, warming permafrost has direct and indirect biophysical impacts, such as landscape changes, hydrological variations, and greenhouse gas releases, as well as socioeconomic impacts such as damage to human settlements and infrastructure, which, in turn, can constrain permafrost distribution and dynamics and the occurrence of permafrost hazards.

Data availability. The dataset is available at https://doi.org/10.5281/zenodo.12157179 (Pan et al., 2024).

Author contributions. SP collected and analyzed the data, prepared the figures, and wrote the paper. PB contributed to the paper and prepared the map of the study sites. MJ assisted with subsequent revisions.

Competing interests. The contact author has declared that none of the authors has any competing interests.

Special issue statement. This article is part of the special issue "Mountain Hydrology and Cryosphere". It is a result of the International Conference on Mountain Hydrology and Cryosphere, Kathmandu and Dhulikhel, Nepal, 9–10 November 2023.

Financial support. Completion of the manuscript was supported by core funds of ICIMOD contributed to by the governments of Afghanistan, Australia, Austria, Bangladesh, Bhutan, China, India, Myanmar, Nepal, Norway, Pakistan, Switzerland, and the United Kingdom.

References

Bajracharya, S. R. and Shrestha, B.: The status of glaciers in the Hindu Kush-Himalayan region, International Centre for Integrated Mountain Development (ICIMOD), https://doi.org/10.53055/ICIMOD.551, 2011.

Baral, P., Allen, S., Steiner, J. F., Gurung, T. R., and McDowell, G.: Climate change impacts and adaptation to permafrost change in High Mountain Asia: a comprehensive review, Environ. Res. Lett., 18, 093005, https://doi.org/10.1088/1748-9326/acf1b4, 2023.

Bian, J., Kou, L., and Zhan, B.: The classification of frozen hazards and protecting approach of permafrost region In Tibetan Plateau permafrost region, The Chinese Journal of Geological Hazard and Control, 22, 129–133, 2011.

Biskaborn, B. K., Smith, S. L., Noetzli, J., Matthes, H., Vieira, G., Streletskiy, D. A., Schoeneich, P., Romanovsky, V. E., Lewkowicz, A. G., Abramov, A., Allard, M., Boike, J., Cable, W. L., Christiansen, H. H., Delaloye, R., Diekmann, B., Drozdov, D., Etzelmüller, B., Grosse, G., Guglielmin, M., Ingeman-Nielsen, T., Isaksen, K., Ishikawa, M., Johansson, M., Johannsson, H., Joo, A., Kaverin, D., Kholodov, A., Konstantinov, P., Kröger, T., Lambiel, C., Lanckman, J. P., Luo, D., Malkova, G., Meiklejohn, I., Moskalenko, N., Oliva, M., Phillips, M., Ramos, M., Sannel, A. B. K., Sergeev, D., Seybold, C., Skryabin, P., Vasiliev, A., Wu, Q., Yoshikawa, K., Zheleznyak, M., and Lantuit, H.: Permafrost is warming at a global scale, Nat. Commun., 10, 264, https://doi.org/10.1038/s41467-018-08240-4, 2019.

Bolch, T., Shea, J. M., Liu, S., Azam, F. M., Gao, Y., Gruber, S., Immerzeel, W. W., Kulkarni, A., Li, H., Tahir, A. A., Zhang, G., and Zhang, Y.: Status and Change of the Cryosphere in the Extended Hindu Kush Himalaya Region, in: The Hindu Kush Himalaya Assessment, edited by: Wester, P., Mishra, A., Mukherji, A., and Shrestha, A., Springer, Cham, https://doi.org/10.1007/978-3-319-92288-1_7, 2019.

Cheng, G. and Jin, H.: Permafrost and groundwater on the Qinghai-Tibet Plateau and in northeast China, Hydrogeol. J., 21, 5–23, https://doi.org/10.1007/s10040-012-0927-2, 2013.

Ci, J., Zhang, Y., Li, C., and Hou, X.: Characteristics and hazards of permafrost from Dushanzi to Qiaoerma in the Tianshan range, Xinjiang, Xinjiang Geology, 24, 71–74, 2006.

Ding, Y., Mu, C., Wu, T., Hu, G., Zou, D., Wang, D., Li, W., and Wu, X.: Increasing cryospheric hazards in a warming climate, 213, 103500, https://doi.org/10.1016/j.earscirev.2020.103500, 2021.

Dobinski, W.: Permafrost, Earth Sci. Rev., 108, 158–169, https://doi.org/10.1016/j.earscirev.2011.06.007, 2011.

Dobiński, W.: Permafrost active layer, Earth Sci. Rev., 208, 103301, https://doi.org/10.1016/j.earscirev.2020.103301, 2020.

Gruber, S. and Haeberli, W.: Mountain Permafrost, in: Permafrost Soils, edited by: Margesin, R., Springer Berlin Heidelberg, Berlin, Heidelberg, 33–44, https://doi.org/10.1007/978-3-540-69371-0_3, 2009.

Haddaway, N. R. and Macura, B.: The role of reporting standards in producing robust literature reviews, Nat. Clim. Change, 8, 444–447, https://doi.org/10.1038/s41558-018-0180-3, 2018.

Harris, C., Arenson, L. U., Christiansen, H. H., Etzelmüller, B., Frauenfelder, R., Gruber, S., Haeberli, W., Hauck, C., Hölzle, M., Humlum, O., Isaksen, K., Kääb, A., Kern-Lütschg, M. A., Lehning, M., Matsuoka, N., Murton, J. B., Nötzli, J., Phillips, M., Ross, N., Seppälä, M., Springman, S. M., and Vonder Mühll, D.: Permafrost and climate in Europe: Monitoring and modelling thermal, geomorphological and geotechnical responses, Earth-Sci. Rev., 92, https://doi.org/10.1016/j.earscirev.2008.12.002, 2009.

Hock, R., Rasul, G., Adler, C., Cáceres, B., Gruber, S., Hirabayashi, Y., Jackson, M., Kääb, A., Kang, S., Kutuzov, S., Milner, A., Molau, U., Morin, S., Orlove, B., and Steltzer, H.: High Mountain Areas, in: IPCC Special Report on the Ocean and Cryosphere in a Changing Climate, edited by: Po¨rtner, H.-O., Roberts, D. C., Masson-Delmotte, V., Zhai, P., Tignor, M., Poloczanska, E., Mintenbeck, K., Alegri´a, A., Nicolai, M., Okem, A., Petzold, J., Rama, B., and Weyer, N. M., Cambridge University Press, Cambridge, UK and New York, NY, USA, 131–202, https://doi.org/10.1017/9781009157964.004, 2019.

Hjort, J., Streletskiy, D., Doré, G., Wu, Q., Bjella, K., and Luoto, M.: Impacts of permafrost degradation on infrastructure, Nat. Rev. Earth Environ., 3, 24–38, https://doi.org/10.1038/s43017-021-00247-8, 2022.

Hu, X. and Yang, Z.-G.: Changes in Soil Microbial Community during Permafrost Collapse Process on the Northeast Qinghai-Tibetan Plateau, China, J. Soil Sci. Plant Nutr., 23, 4157–4167, https://doi.org/10.1007/s42729-023-01332-1, 2023.

James, K. L., Randall, N. P., and Haddaway, N. R.: A methodology for systematic mapping in environmental sciences, Environ. Evid., 5, 7, https://doi.org/10.1186/s13750-016-0059-6, 2016.

Li, D., Lu, X., Walling, D. E., Zhang, T., Steiner, J. F., Wasson, R. J., Harrison, S., Nepal, S., Nie, Y., Immerzeel, W. W., Shugar, D. H., Koppes, M., Lane, S., Zeng, Z., Sun, X., Yegorov, A., and Bolch, T.: High Mountain Asia hydropower systems threatened by climate-driven landscape instability, Nat. Geosci., 15, 520–530, https://doi.org/10.1038/s41561-022-00953-y, 2022.

Liu, C., Li, T., and Lin, Y.: Study on environment geology impact assessment for highway rebuilding in westward altiplano region, Environmental Protection in Transportation, 24, 8–11, 2003.

Liu, Z.: Distribution characteristics of permafrost development along Qinghai-Tibet Railway, Railway Investigation and Surveying, 24, 78–82, 2008.

Luo, J., Niu, F., Lin, Z., Liu, M., and Yin, G.: Development of thawing hazards and thermal influence on permafrost along Qinghai-Tibet enginering corridor, Journal of Engineering Geology, 22, 326–333, 2014.

Ma, D., Zhang, J., Wang, M., Zhao, J., and Gao, W.: Characteristics and hazards of permafrost along the Xinjiang-Tibetan Highway, Xinjiang, J. Mt. Sci., 22, 554–561, 2004.

Ma, W., Niu, F., and Mu, Y.: Basic research on the major permafrost projects in the Qinghai-Tibet plateau, Adv. Earth Sci., 27, 185–191, 2012.

Mu, M.: Characteristics and impact factors of carbon release from thaw slump landforms in Qilian, Master dissertation, Lanzhou University, Lanzhou, https://ir.lzu.edu.cn/handle/262010/460095 (last access: 6 August 2024), 2021.

Pan, S., Baral, P., and Jackson, M.: Sup_Assessment of permafrost related hazards in China based on Chinese literature, Zenodo [data set], https://doi.org/10.5281/zenodo.12157179, 2024.

Pang, W., Chang, G., Gou, H., and Wang, Y.: Characteristics of the permafrost from Qilian to Mole of the Highway 204 in Qinghai Province, J. Glaciol. Geocryol., 41, 93–99, 2019.

Patton, A. I., Rathburn, S. L., and Capps, D. M.: Landslide response to climate change in permafrost regions, Geomorphology, 340, 116–128, https://doi.org/10.1016/j.geomorph.2019.04.029, 2019.

Qinglai, C.: Engineering geologial research on the permafrost in high lattitudes area and its impact on pipeline construction, PhD dissertation, Chinese Academy of Geological Sciences, Beijing, https://kns.cnki.net/KCMS/detail/detail.aspx?dbname=CDFD0911&filename=2007213368.nh (last access: 6 August 2024), 2007.

Ran, Y., Li, X., Cheng, G., Zhang, T., Wu, Q., Jin, H., and Jin, R.: Distribution of Permafrost in China: An Overview of Existing Permafrost Maps, Permafrost Periglac., 23, 322–333, https://doi.org/10.1002/ppp.1756, 2012.

Ran, Y., Cheng, G., Dong, Y., Hjort, J., Lovecraft, A. L., Kang, S., Tan, M., and Li, X.: Permafrost degradation increases risk and large future costs of infrastructure on the Third Pole, Commun. Earth Environ., 3, 328, https://doi.org/10.1038/s43247-022-00568-6, 2022.

Shang, J.: Study on change of thaw slump based on remote sensing Over the Eboling Ridge in permafrost region of Qilian Mountains, Master dissertation, Lanzhou University, Lanzhou, https://kns.cnki.net/KCMS/detail/detail.aspx?dbname=CMFD201802&filename=1018979558.nh (last access: 6 August 2024), 2018.

Tananaev, N. and Lotsari, E.: Defrosting northern catchments: Fluvial effects of permafrost degradation, 228, 103996, https://doi.org/10.1016/j.earscirev.2022.103996, 2022.

Wang, F., Li, G., Ma, W., Wu, Q., Serban, M., Vera, S., Alexandr, F., Jiang, N., and Wang, B.: Pipeline–permafrost interaction monitoring system along the China–Russia crude oil pipeline, Eng. Geol., 254, 113–125, https://doi.org/10.1016/j.enggeo.2019.03.013, 2019.

Wang, J., Wang, S., and Qiu, G.: Permafrost along the Qinghai-Xizang Highway, Acta Geogr. Sin., 34, 18–32, 1979.

Wang, X., Ran, Y., Pang, G., Chen, D., Su, B., Chen, R., Li, X., Chen, H. W., Yang, M., Gou, X., Jorgenson, M. T., Aalto, J., Li, R., Peng, X., Wu, T., Clow, G. D., Wan, G., Wu, X., and Luo, D.: Contrasting characteristics, changes, and linkages of permafrost between the Arctic and the Third Pole, Earth-Sci. Rev., 230, 104042, https://doi.org/10.1016/j.earscirev.2022.104042, 2022.

Wu, Q., Zhang, T., and Liu, Y.: Permafrost temperatures and thickness on the Qinghai-Tibet Plateau, Global Planet. Change, 72, 32–38, https://doi.org/10.1016/j.gloplacha.2010.03.001, 2010.

Xu, X., Zhang, Z., and Wu, Q.: Simulation of permafrost changes on the Qinghai–Tibet Plateau, China, over the past three decades, Int. J. Digit. Earth, 10, 522–538, https://doi.org/10.1080/17538947.2016.1237571, 2017.

Yao, T., Xue, Y., Chen, D., Chen, F., Thompson, L., Cui, P., Koike, T., Lau, W. K.-M., Lettenmaier, D., Mosbrugger, V., Zhang, R., Xu, B., Dozier, J., Gillespie, T., Gu, Y., Kang, S., Piao, S., Sugimoto, S., Ueno, K., Wang, L., Wang, W., Zhang, F., Sheng, Y., Guo, W., Ailikun, Yang, X., Ma, Y., Shen, S. S. P., Su, Z., Chen, F., Liang, S., Liu, Y., Singh, V. P., Yang, K., Yang, D., Zhao, X., Qian, Y., Zhang, Y., and Li, Q.: Recent Third Pole's Rapid Warming Accompanies Cryospheric Melt and Water Cycle Intensification and Interactions between Monsoon and Environment: Multidisciplinary Approach with Observations, Modeling, and Analysis, B. Am. Meteor. Soc., 100, 423–444, https://doi.org/10.1175/BAMS-D-17-0057.1, 2019.

Yin, G., Niu, F., Lin, Z., Luo, J., Liu, M., and Li, A.: The distribution characteristics of permafrost along the Qinghai-Tibet Railway and their response to environmental change, J. Glaciol. Geocryol., 36, 772–781, 2014.

Zhang, T.: Historical overview of permafrost studies in China, Phys. Geogr., 26, 279–298, https://doi.org/10.2747/0272-3646.26.4.279, 2005.

Zhou, Y., Guo, D., Qiu, G., Cheng, G., and Li, S.: Geocryology in China, 1st edn., Science Press, Beijing, ISBN 9787030082855, 2000.

Zhu, F.: Mechanism of frost heave and thaw settlement of subgrade and stability evaluation of subgrade in alpine permafrost regions, Master dissertation, Chang'an University, Xi'an, https://kns.cnki.net/KCMS/detail/detail.aspx?dbname=CMFD201801&filename=1017870452.nh (last access: 6 August 2024), 2017.

AUTHOR INDEX